ACID RAIN

ECONOMIC ASSESSMENT

ENVIRONMENTAL SCIENCE RESEARCH

Recent Volumes in this Series

Volume 27 – SHORT-TERM BIOASSAYS IN THE ANALYSIS OF COMPLEX ENVIRONMENTAL MIXTURES III
Edited by Michael D. Waters, Shahbeg S. Sandhu, Joellen Lewtas, Larry Claxton, Neil Chernoff, and Stephen Nesnow

Volume 28 – UTILIZATION OF MAMMALIAN SPECIFIC LOCUS STUDIES IN HAZARD EVALUATION AND ESTIMATION OF GENETIC RISK
Edited by Frederick J. de Serres and William Sheridan

Volume 29 – APPLICATION OF BIOLOGICAL MARKERS TO CARCINOGEN TESTING
Edited by Harry A. Milman and Stewart Sell

Volume 30 – INDIVIDUAL SUSCEPTIBILITY TO GENOTOXIC AGENTS IN THE HUMAN POPULATION
Edited by Frederick J. de Serres and Ronald W. Pero

Volume 31 – MUTATION, CANCER, AND MALFORMATION
Edited by Ernest H. Y. Chu and Walderico M. Generoso

Volume 32 – SHORT-TERM BIOASSAYS IN THE ANALYSIS OF COMPLEX ENVIRONMENTAL MIXTURES IV
Edited by Michael D. Waters, Shahbeg S. Sandhu, Joellen Lewtas, Larry Claxton, Gary Strauss, and Stephen Nesnow

Volume 33 – ACID RAIN: Economic Assessment
Edited by Paulette Mandelbaum

A Continuation Order Plan is available for this series. A continuation order will bring delivery of each new volume immediately upon publication. Volumes are billed only upon actual shipment. For further information please contact the publisher.

ACID RAIN

ECONOMIC ASSESSMENT

Edited by
Paulette Mandelbaum
Chemical Engineering and Policy Analyses (CEPA)
Rochester, New York

Technical Editor
Carole Beal
Acid Rain Information Clearinghouse
Rochester, New York

PLENUM PRESS • NEW YORK AND LONDON

Library of Congress Cataloging in Publication Data

Conference on Acid Rain (1984: Washington, D.C.)
Acid rain.

(Environmental science research; v. 33)
"Proceedings of a Conference on Acid Rain sponsored by the Acid Rain Information Clearinghouse, a project of the Center for Environmental Information, held December 4–6, 1984, in Washington, D.C."—T.p. verso.
Bibliography: p.
Includes index.
1. Acid rain—Economic aspects—United States—Congresses. 2. Acid rain—Government policy—United States—Congresses. I. Mandelbaum, Paulette A. II. Acid Rain Information Clearinghouse. (Washington, D.C.) III. Title. IV. Series.
TD196.A25C66 1984 363.7′386 85-17021
ISBN-13:978-1-4615-8355-4 e-ISBN-13:978-1-4615-8353-0
DOI: 10.1007/978-1-4615-8353-0

Proceedings of a conference on Acid Rain sponsored by the Acid Rain Information Clearinghouse, a project of the Center for Environmental Information, held December 4–6, 1984, in Washington, D.C.

The Center for Environmental Information Inc. gratefully acknowledges funding provided by the following sources in support of the publication of these proceedings:

Association for the Protection of the Adirondacks
Edison Electric Institute
Federal Government of Canada
National Acid Precipitation Assessment Program
United States Environmental Protection Agency

Softcover reprint of the hardcover 1st edition 1985

A Division of Plenum Publishing Corporation
233 Spring Street, New York, N.Y. 10013

ABOUT THE CENTER FOR ENVIRONMENTAL INFORMATION

The Acid Rain Information Clearinghouse is a project of the Center for Environmental Information Inc., 33 South Washington Street, Rochester, New York 14608. CEI, a private, nonprofit organization established in 1974, provides information through its publications, access to computerized information sources, educational programs, conferences and library. *As a matter of policy, CEI does not take positions on environmental issues.*

CONFERENCE ADVISORY COMMITTEE

Richard Chastain
Southern Company Services
Thomas D. Crocker, Ph.D.
Department of Economics, University of Wyoming
William Hogan, Ph.D.
Energy and Environmental Policy Center, Harvard University
Steve Howard
National Wildlife Federation
Jay S. Jacobson, Ph.D.
Boyce Thompson Institute for Plant Research
Lester B. Lave, Ph.D.
Graduate School of Industrial Administration, Carnegie-Mellon University
Paul W. MacAvoy, Ph.D.
Graduate School of Management, University of Rochester
Paul Stolpman
United States Environmental Protection Agency
Greg Wetstone
Health and Environment Subcommittee, House of Representatives, United States Congress

CONFERENCE MODERATORS

William Hogan, Ph.D.
Energy and Environmental Policy Center, Harvard University
Paul A. Miller, Ph.D.
Department of Science and Humanities, Rochester Institute of Technology

PREFACE

This volume, Proceedings of the Conference ACID RAIN: Economic Assessment, is meant to present the areas of agreement which economists have established and the uncertainties which they have discovered in their attempts to use the methodology of economics to better understand the nature of the acid rain issue.

Scientific articles about acid rain initially appeared in 1972. The public turned its attention to the issue in the mid-1970s. In April 1979, the first acid rain bill was introduced in the Senate, authored by New York's Senator Daniel P. Moynihan. The bill sought to establish a federal research program dedicated to filling the gaps in understanding of the phenomena of long-range transport of air pollutants and their environmental, health and economic impacts. The bill was passed into law in 1980. Since then, tens of bills have been proposed to control emissions of SO_2 and NO_x, thought to be the precursors of acid rain. And yet, in contrast with the pattern set by the majority of environmental issues, where legislation followed very quickly on the heels of public anxiety and involvement, by July 1985 not a single federal acid rain control bill had been passed. The reasons for the log jam in Congress and the federal government have been analyzed in great detail, but add up to a single fact: the traditional coalitions which delivered national congressional majorities for pro-environmental laws, and in particular for the Clean Air Act of 1977, had dissolved because some members of the coalition, such as the northeastern states, stood to gain a great deal from acid rain control bills, while others such as the coal industry and coal miners of the Midwest, stood to lose a great deal. The politics of acid rain were further complicated in spring of 1983, when efforts were undertaken to win midwestern support by levying a tax across the nation. At that juncture, the western states and Hawaii were brought into the foray and generally took stands opposing the tax.

Quite apart from the questions of regionalism and fragmentation of the traditional clean air lobby, there was a firm feeling from mid-1981 in the Office of Management and Budget and the White House, that the costs of acid rain controls would add up to billions of dollars although at the same time, the benefits were showing up as millions of dollars. The Reagan Administration would commit funds to research, but would take no steps toward initiating an acid rain control strategy.

By early 1984, while the congressional politics of acid rain were caught in a stalemate, there was a realization at the Acid Rain Information Clearinghouse that as the scientific community slowly generated results and developed a more complete understanding of the causes and effects of acid deposition, a need could arise to make decisions in the absence of scientific certainty, based on some sort of socio-economic analysis.

It was to create a forum for discussion of what decisionmakers, constrained by the political economy of the acid rain issue, could do with the uncertain scientific information as it emerged, that ARIC's December conference was held. Two sorts of questions were asked of the economists invited to speak. One was methodological, and concerned the economist's ability to assess or estimate factors as complex as the national or regional costs and benefits of acid rain controls. The other touched on societal values, and asked economists to address the question frequently raised by the environmental community, that is, whether or not efficiency ought to play a decisive role in environmental debate. Because of the great uncertainties which accompany current understanding of the physics and chemistry of acid rain and the effects of acid deposition, as well as uncertainties associated with some components of the economic analysis, decision-theorists were asked to discuss methods of risk assessment and decisionmaking in the absence of scientific certainty.

The first three sessions of the conference, and the guest addresses by the OECD's Ian Torrens and the United Mine Workers' President Richard Trumka, dealt with the economic problems noted above. One conference session and the guest addresses by New York's Commissioner of Environmental Conservation, Henry Williams, and EPA's Assistant Administrator for Air and Radiation, Charles Elkins, dealt with modes of decisionmaking in the face of uncertainty. In addition, because scientists, economists and policy analysts are all working on acid rain problems, and because each group is impatiently awaiting results or analysis from one or another of the fields, a session was built around discussion of the difficulties which scientists, economists, policy analysts and decisionmakers encounter, both in communicating their needs to one another and in attempting to obtain the research results of which they are in need. Allen Kneese, a leading environmental economist and Senior Fellow at Resources for the Future, reviewed each of the papers before the conference. In concluding remarks, he assessed the state of the art in economic methodologies and its application to the acid rain issue.

The publication of these Proceedings is the result of the cooperation of a great number of people. Thanks are due to the 26 speakers and respondents, who provided us with their articles in time and were willing to follow tedious requests to review their papers as we moved from one format to another. We are indebted to Ronald Dwight and the Carnegie Endowment for International Peace, for co-sponsoring Ian Torrens' keynote address to the conference in their Face-to-Face series. The talk and the informal discussion which followed set a tone of excellence and openness for the conference.

Warmest thanks are due to A. Lee Nesslage, of the Center for Environmental Information, who coordinated the conference arrangements and was always willing to make changes in her well thought-out plans in order to accommodate the needs of the program, and to Frederick W. Stoss, Manager of Information Services at the Acid Rain Information Clearinghouse, for providing general background information and documentation for the conference.

We are grateful to a number of people who assisted in the preparation of the manuscript. Thanks are due to William R. Wagner, Manager of Communications of the Center for Environmental Information, for his work in arranging the figures and tables in the book, to Mrs. William H. Wagner, for faithfully transcribing the talks which were presented and the discussions which took place at the conference, to Steve

Berendt of COMPUTYPE, who typed the bulk of the manuscript in record time and with true proficiency, and to Shelley Gordon and Eileen Williams of the Manuscript Division at the University of Rochester Computer Center, for their willingness to experiment with the transfer of the manuscript from one format to another, and for turning the manuscript into camera-ready copy. Great thanks are due to Ann Held, Dianne Piccirilli and Wanda Versprille, of Chemical Engineering and Policy Analyses (CEPA) of Rochester, New York, who over the past year and one-half have canvassed the speakers and coordinated materials at each juncture of the preparation of the conference and proceedings.

Finally, we are most thankful to Carole Beal of the Center for Environmental Information, who shepherded the manuscript from the first to the third and final draft, spending countless hours to produce a consistent and error-free text, and who carefully indexed the entire book.

Paulette Mandelbaum
June 30, 1985
Editor

CONTENTS

CONCLUDING REMARKS

DINNER SPEAKERS AND GREETINGS

WELCOMING REMARKS

Elizabeth Thorndike

Executive Director
Center for Environmental Information Inc.
33 South Washington Street
Rochester, NY 14608

On behalf of the Center for Environmental Information and the Acid Rain Information Clearinghouse, I want to welcome you to this conference. For those of you unfamiliar with the sponsor, let me offer a brief synopsis. The Center for Environmental Information Inc., founded in 1974, is an independent, nonprofit organization, located in Rochester, New York. The Acid Rain Information Clearinghouse is a project of the Center.

We do not lobby, we do not litigate, we do not do research, and as a matter of policy, we do not take positions on issues. That policy is reflected in the diverse composition of our Board of Directors and of our Advisory Councils including that of the Acid Rain Information Clearinghouse.

We do gather, organize and disseminate information on the environment by means of publications, such as the *Acid Precipitation Digest*, response to inquiries, and through seminars, workshops and conferences such as this one. Providing accurate data and encouraging constructive dialogue describe the thrust of our organization.

This gathering is the first national conference to deal entirely with economic assessment of the acid rain issue. It is timely. As your registration folder indicates, we have identified over 180 studies, reports, and articles dealing with the economic aspects of acid rain, each with differing assumptions, interpretations and numbers. Is it any wonder that our decisionmakers are confused?

This conference is not intended to recommend a solution in dollars and cents. Our goals are to take first steps in sorting out and understanding the available information by:

- identifying the ways in which economics can aid in resolution of the acid rain debate;
- examining the assumptions underlying economic assessment; and
- evaluating whether cost-benefit analysis is valid when applied to the acid rain issue.

Our economic well-being is tied to the health of our natural resources. We hope the outcome of this conference dialogue will move us closer to a solution which accommodates both the environmental integrity and economic viability of our country.

ACID RAIN: AN INTERNATIONAL PERSPECTIVE

ACID RAIN -- AN INTERNATIONAL PERSPECTIVE

Ian M. Torrens*

Head of the Resources and Energy Division
Environmental Directorate
Organization for Economic Cooperation & Development
2, rue André Pascal
75775 Paris Cedex 16
France

Keynote Address Co-Sponsored by the Carnegie Endowment for International Peace and the American Foreign Service Association

1950-1980: GROWING INTERNATIONAL AWARENESS

I have been asked this evening to broaden your perspective, if in fact it should need broadening, on the historical development and present situation as regards the international debate on acid rain. In fact, acid rain was a controversial issue in Europe when many of those present in this room were in primary school. During the 1950s and 1960s, a network of measuring stations in Scandinavia indicated gradually increasing acidity of rainfall, and observations in rivers and lakes pointed to many instances of decreasing fish population.

These Scandinavian observations were presented in 1969 to an OECD meeting, where it was decided to extend the measurement to cover a substantial area of OECD Europe. The OECD program, launched in 1972 and completed in 1977, confirmed that sulfur compounds do travel long distances in the atmosphere, and showed that the air quality in any one European country is measurably affected by emissions from other European countries. Thus it was clear that no one country going it alone could hope to solve the problems due to air pollution, and that some sort of international cooperative program would be necessary.

The problem, however, was wider than simply the OECD. Our program showed that air pollutants have no respect for national frontiers between OECD countries. Neither, however, do they respect geopolitical regional boundaries, as between eastern and western European countries. For a control program to be really effective, therefore, it would have to encompass eastern Europe too.

*The opinions expressed in this paper are those of the author and do not necessarily represent the views of the OECD or of the governments of its member countries.

Consequently, the United Nations Economic Commission for Europe, based in Geneva, took up the challenge, establishing a Cooperative Programme for Monitoring and Evaluation of the Long-Range Transmission of Air Pollutants in Europe (EMEP), and developing a Convention on Long-Range Transboundary Air Pollution, which was signed by 34 ECE countries (including the United States and Canada) in 1979.

While the ECE Convention was being negotiated in Geneva, interest and concern about acid rain was growing rapidly in North America, and a series of studies in both countries led to intergovernmental discussions which resulted in the signing by the governments of Canada and the United States in 1980 of a Memorandum of Intent to develop a bilateral agreement on transboundary air pollution. I do not need to elaborate on the North American part of the international debate, given the present audience's familiarity with it. Canadian calls for prompt action to reduce air pollution have so far met with a cautious response from the U.S. Administration, accompanied, however, by a major research effort to clarify some of the key issues related to acid deposition.

THE EARLY 1980s: A NEW IMPETUS

Continuing this historical survey, the scene now shifts in the early 1980s to the Federal Republic of Germany, a country which had up to that time been categorically opposed to any major increase in controls on emissions of acid gases. It was discovered relatively suddenly (i.e. over a period of a year or two) that major forest damage was in progress in several parts of the country, such as the Black Forest and the forests of Bavaria. Each year since 1981 has seen a remarkable increase in the estimates of the forest area affected (from 8 percent in 1982, to 34 percent in 1983, to 54 percent as a latest estimate). Air pollution is considered by the German government to be a prime suspect in the mystery of tree death, though categorical proof is not yet available, and other possible causes have been suggested (e.g. effects of periodic drought in recent years, and forestry management practices in Germany). However, der Tannenbaum, the fir tree, in German culture is something special, and there is a strong political push to take measures rapidly to stop spreading damage to forests. The politics of the environment in the Federal Republic of Germany are of course affected by the growing presence of the "Green" Party in federal and state parliaments there. One consequence of this combination of circumstances is a keen environmental awareness in all other major political parties.

Thus Germany's position on acid rain turned around 180° during the period 1982-83. Internally, a set of severe emission regulations for large combustion installations were promulgated in 1983, which affect both new and existing plants. Existing plants with a projected operating lifetime of more than 30,000 hours will have to be retrofitted with pollution control equipment adequate to meet the stated emission limits by 1988 at the latest. All plants not meeting these standards must be retired by 1993. It is estimated that 80 percent of German hard coal and lignite electricity generating capacity will be retrofitted with flue gas desulfurization (FGD) over the next four years. In addition, regulations requiring the use of flue gas denitrification in large installations may be promulgated in the Federal Republic of Germany in the near future.

More recently, the German government turned its attention to mobile sources of air pollution. By and large, emission limits for vehicles have been less strict in

Europe than in either the United States or Japan. For example, catalytic converters have not to date been necessary in European cars. Germany and some other European countries (principally Sweden and Switzerland) are leading a political drive to bring in emission standards for vehicles in Europe which would parallel U.S. standards.

What have vehicles got to do with acid rain, one might ask? They emit about 50 percent of the total national emissions of nitrogen oxides, which make up about 30 percent of the acidity in rainfall. But the reasons are more complex. Motor vehicles also emit hydrocarbons, which interact with nitrogen oxides in the atmosphere to form ozone. Ozone is now suspected of being among the possible causes of forest damage, hence the interest in reducing emissions of its precursors from vehicles as well as stationary sources.

The present situation with regard to reducing motor vehicle emissions in Europe is exceedingly complex, with a considerable number of countries involved. Moreover, this issue has become co-mingled with another major political issue, that of reducing or eliminating lead in gasoline via the principal technology currently available for emission reduction, namely the catalytic converter.

The European Commission in Brussels, at the initiative of the German government, has adopted a directive on the availability of unleaded gasoline in EC countries. But important issues as regards trade in motor vehicles between European countries have been raised, and positions of major countries and motor vehicle manufacturers in the present debate differ quite a lot as to the best way to proceed.

I would not like to prejudge the outcome of this debate, other than to say I think it probable that cars in Europe will have to face stricter emission standards, perhaps equivalent to those now applicable in the United States, by a given year in the not-too-distant future, perhaps sometime in the 1988-95 range.

Returning for a moment to stationary installations, and in particular sulfur dioxide control, the Nordic countries, at the first meeting of the Executive Body for the ECE Convention on Long-Range Transboundary Air Pollution, held in Geneva in June 1983, proposed that signatories to the Convention should agree to reduce their national emissions of sulfur dioxide by 30 percent by 1993 compared to 1980 emission levels. Though this was not accepted by the whole Executive Body, it formed the nucleus of what has come to be known as the "Thirty Percent Club", whose entry fee is a national commitment to reduce SO_2 emissions by that amount before 1993. Membership of the "Club" has grown significantly over the past 18 months, catalyzed by several major international meetings. One such meeting held in Ottawa last spring (limited to members of the "Club") brought on board a few more countries.

The government of the Federal Republic of Germany, consistent with its strong internal environmental stance, has taken a very active position vis-a -vis the international community. It arranged a major Multilateral Conference on the Environment in Munich in June of this year, inviting all ECE member countries. Many were represented at Ministerial level, and a set of resolutions were adopted to be transmitted to the Executive Body in Geneva. The principal of these was that the Conference recommended that an agreement on a specific reduction in sulfur emissions be negotiated in the ECE. Accordingly, at the second meeting of the

Executive Body in September this year, a negotiating group was set up to draft a Protocol for such an agreement. These negotiations are in progress and it is hoped to complete them next spring.

At the Munich Conference several more countries joined the "Thirty Percent Club". It contains a total of twenty countries, including a number of eastern European ones (which, however, agree to reduce not national emissions but transboundary flows of SO_2 by 30 percent). The United States has not to date expressed its willingness to join the "Club", on the basis that future reductions of SO_2 emissions will occur according to present policies as old plants are replaced by new ones, and that a policy decision to take further measures on top of this should await the outcome of the major research program on acid rain currently under way.

ACID RAIN AND AIR POLLUTION: CLEARING UP SOME CONFUSION

The term "acid rain" has become shorthand for a widening range of perceived environmental problems stemming from air pollution, and in common with many such over-simplifications, its use, or misuse, often leads to inaccuracies and confusion. This can be quite counterproductive in a field where the complexities render the most knowledgeable attempt to describe the causes and effects fraught with uncertainty.

It would be advisable, therefore, to set the bounds as clearly as possible at the outset. In fact, acid rain is only one aspect of a wider set of environmental problems which owe their origin, in whole or in part, to the range of air pollutants emitted by various industrial or other economic activities. The more correct term for acid rain is acid deposition, which includes both wet and dry forms. But in order to understand better the effects of acid deposition, and to take measures to remedy them, we need to widen our horizons still further to consider air pollution in general, since some of the effects seem to be related to the quality of the airshed rather than to particular substances which are deposited in a specific place.

The major air pollutants implicated in the types of phenomenon attributed to acid deposition are sulfur oxides (SO_x), nitrogen oxides (NO_x), hydrocarbons (HC) and particulate matter (PM). These pollutants are emitted in various proportions from both stationary sources and vehicles. They are subject to interaction with each other and with other atmospheric substances, and can travel considerable distances depending on the conditions of their emission and on climatic factors. Any strategies to address the problems of air pollution should be developed with the objective of covering all sources of these pollutants, both stationary and mobile.

AIR POLLUTION TRENDS AND REGULATORY APPROACHES

A point which is often overlooked in the present debate is that in many countries there has been a degree of success over the past decade or so in controlling these pollutants. For example, in the United States total SO_2 emissions fell by about a quarter from their peak in 1972 until 1982, even though the use of fossil fuels and industrial processes responsible for these emissions grew substantially over this period (Figure 1). The figure also shows how SO_2 emissions would have increased if pollution controls in the United States had remained at 1970 levels throughout the period.

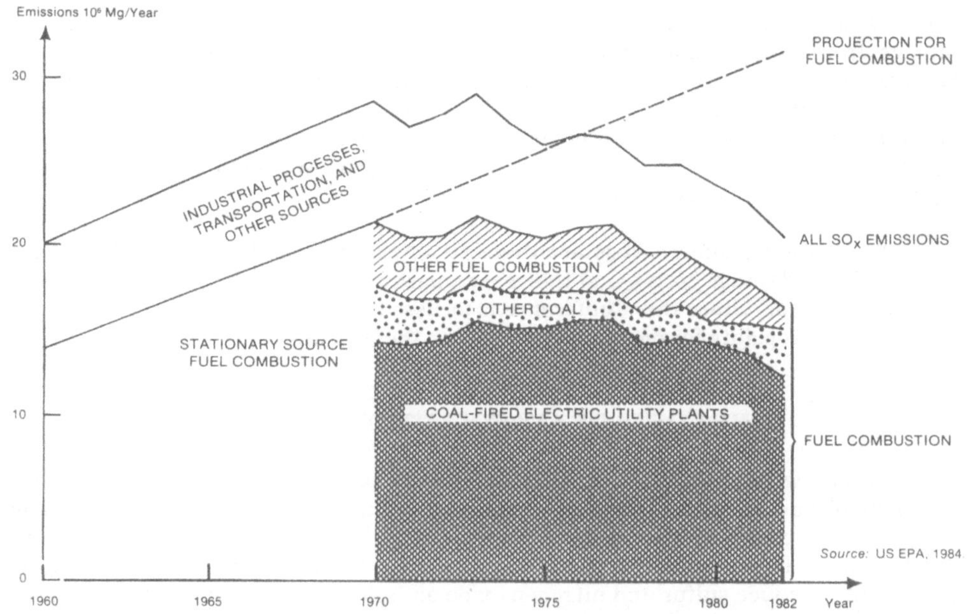

Figure 1. Estimated U.S. Emissions of SO_x by Source Type

This same pattern is true for a number of other OECD countries, and for some of them, present control regulations will lead to future reductions in emissions as older, more polluting sources are replaced by newer and cleaner ones. This does not, however, imply that nothing more needs to be done to reduce acidifying emissions. The crux of the present debate is the question of how much more, when, and at what cost?

All OECD countries exercise some form of control over airborne emissions resulting from combustion of fossil fuels, particularly for electricity generation. This control may be either direct, by applying prescribed emission limits, or indirect, by regulating the quality or type of fuel used. The OECD has recently issued a report containing a compendium of emission standards for major air pollutants from energy facilities (OECD, 1984). It shows the very considerable degree of diversity both in approach to regulation and, where emission limits are applied, in the numerical level at which these limits are set.

THE ECONOMIC ISSUES

The OECD, which had carried out pioneering work in the field of acid rain in the 1970s, did not of course opt out when negotiations moved to ECE in Geneva. We still had (and have) a lot to contribute to the understanding of the air pollution problems and to their control.

As befits an organization whose basic concern is the economic and social development of its twenty-four member countries, the OECD has tended to approach the question of damage from environmental pollution and policies to reduce that damage, from the economic perspective. Since the current estimates of the size of the damage caused by air pollution in Europe and North America are matched only by the current estimates of the size of the control costs required to reduce emissions of the pollutants responsible, with both running into billions (not millions) of dollars annually, the economics of the problem and of the proposed solutions are bound to figure strongly in the policy debate.

The main economic issues are the costs of control, the benefits of control (or environmental damage costs), and the vexed question of who should bear the costs of additional control. Since this Conference is focussed on economic assessment, I think it would be appropriate for me to touch briefly on these issues.

THE COSTS OF AIR POLLUTION CONTROL

Recent work in OECD on coal pollution abatement (OECD, 1983 a,b) has highlighted the difficulties involved in estimating the costs of pollution control, and particularly in comparing estimates across countries. There are many uncertainties. But this work has established (and other recent estimates concur) that installing equipment to reduce sulfur and nitrogen oxide emissions from a new coal-fired power plant by 90 percent, and particulate emissions by more than 99 percent, would increase the cost of electricity generation from that plant by some 15-25 percent compared to an uncontrolled facility. This includes flue gas denitrification, to date practiced only in Japan on power plants. Installing pollution control equipment in an existing power plant would be somewhat more expensive, the amount depending on specific characteristics of the plant and the site.

If the estimation of pollution control costs is complex for a single plant, a much higher degree of uncertainty applies to estimates of the cost of acid rain control strategies at the national or international scale. Among the sources of uncertainty, even given the constancy of environmental regulations, are the difficulty in assessing the future electricity demand; the rate at which new generating capacity comes on stream and old capacity is phased out; and the difficulty in predicting when technological improvements will be available for commercialization and at what cost.

Notwithstanding these difficulties, overall cost estimates have been made for the costs of acid rain control strategies in Europe and the United States. These include the installation of pollution control equipment on both new and existing power plants and industrial installations. Table 1 gives you an idea of the order of magnitude.

These are, of course, large numbers. But we should try to get them into the perspective of what they would mean for the electricity consumer, for example. In the German case, the SO_2 reduction of Table 1 is estimated to add up to one pfennig per kilowatt-hour to the electricity price to consumers (or approximately 6 percent). The United Kingdom Central Electricity Generating Board has estimated that retrofitting FGD to existing coal-fired power plants to achieve a 50 percent reduction might add, over a period of more than a decade, about 5-6 percent in total to the average cost of electricity generated in the United Kingdom, and considerably less to consumers' electricity bills. These are not negligible increases, but they are far from catastrophic.

Table 1

Costs of Strategies to Control Air Pollutants

Country or Group	Air Pollutant	Emission Source (Stationary)	Reduction	By Year	Annual Cost	Source
European Community	SO_2	all	10-13 M tonnes (53-77%)	2000	$4.6-6.7 bn(1)(2)	ERL, 1983
	NO_x	all	50%	2000	$0.4 bn(1)	ERL, 1983
Federal Republic of Germany	SO_2	all	1.6 M tonnes (50%)	1993	DM3.3 bn	Federal Parliament, 1984
United Kingdom	SO_2	power plants	1.5 M tons (50%)	1995+	£175 m(3) (average)	Royal Commission, 1984 UK House of Commons, 1984
United States	SO_2	power plants	10 M tons (75%)	1995+	$4.2-5.3(1)	ICF, 1984
United States	SO_2	power plants	"	"	$5.2-9.5(1)	EPRI, 1983

(1) 1982 prices.
(2) Includes retrofitting FGD on 70% of large coal and lignite boilers over 25 MW.
(3) 1983 prices. Includes cost of replacement of output capacity lost through retrofitting of FGD. This accounts for about 25% of total capital costs of £1990 million sterling.

This brief overview of the costs side has not addressed the costs of control for mobile sources of air pollution. In fact, as I mentioned earlier, a very different situation prevails in the different regions of OECD, and it would be difficult to come up with any comparable numbers for overall control costs.

According to the U.S. Department of Commerce (1981), all controls of NO_x, CO, HC, and PM from mobile sources cost $16.5 billion annually in 1981 (1981 dollars). These costs are the total for the approximately 136 million light and heavy duty vehicles in the United States in that year, for an average of $121 per vehicle per year. (That figure, which seems high, covers all costs of the mobile source air pollution program, including program administration and enforcement, fuel maintenance, and other costs added by controls, but not R&D). With this sum, the U.S. has achieved reduction of over 90 percent of emissions from auto gasoline engines, and has reduced diesel emissions as well.

BENEFITS OF AIR POLLUTION CONTROL

No analysis would be complete without referring to the costs of *not* controlling these air pollutants, or of doing so to an inadequate degree. The issue of benefits of pollution control is a very complex one, because of the great diversity of the field of environmental damage needing to be covered; the uncertainties in linking air pollutant emissions or even ambient concentrations with damages produced, particularly in a quantitative way; the difficulties and value judgments involved in expressing these damages in monetary terms; and the intangible nature of some of the effects, such as damage to historical monuments, which makes conversion to monetary terms impossible in such cases.

To cite the figures from some of the major benefits studies which have been carried out, including one by the OECD (OECD, 1981) and a more recent one carried out for the European Community by Environmental Resources Limited (ERL, 1983), would require me to add so many qualifying cautionary statements that I would far exceed the time available for me today. Suffice it to say that an increasing number of serious studies suggest that the costs and benefits of substantial reductions in pollutant emissions could be of a similar order of magnitude, and that a significant fraction of the benefits do not lend themselves to easy quantification.

A second economic factor on the benefits side, which is not often taken into account in the policy debate, is the stimulative impact which stricter environmental standards can have on economic activity and on technological development. R&D and manufacturing pollution control equipment contribute to our economic progress, and in particular to GDP. They can also add to levels of employment in sectors which are at present working well below capacity. Finally, they can result in lowering both environmental control costs and other costs through technological advances such as the development of more fuel-efficient automobiles.

WHO BEARS THE COSTS?

This is one of the most complex and sensitive issues, for three basic reasons:

- First, the magnitude of the costs of control or damage measured, as mentioned earlier, in billions of dollars;

- Second, the fact that, in most cases, those who would be asked to bear the cost of pollutant emission reductions are not the same as those who presently bear the costs of environmental damage and often are not even citizens of the same country;

- Third, because of the time it may take to slow down and reverse some of the damage processes, the costs of control and of environmental damage may have to be paid simultaneously for a number of years, perhaps even for decades.

Underlying much of the environmental work in the OECD is the "Polluter Pays Principle" (PPP), which says that the costs of preventing pollution should be borne by the polluter and reflected, to the extent that the market permits, in the cost of the polluter's products.

It is sometimes perceived to be politically difficult to reflect the PPP in policies to reduce air pollution, because of the magnitude of the costs this could place on a segment of the population, particularly those who use electricity generated from power plants which are required to install pollution control equipment. This reasoning is familiar in the OECD, having been used consistently over the past decade by all those municipalities, manufacturers or other industrial polluters to which this principle was likely to apply. However, for sound economic reasons, the OECD will continue to support the application of PPP, unless and until a very convincing economic argument can be made for an alternative method of paying for pollution prevention.

THE POLICY DILEMMA

Policymakers are therefore faced with something of a dilemma, depicted in simplified terms in Figure 2. Basically, the choice is to act now, or to wait until more is known, being conscious of the implications of each of the two choices.

The key question facing policymakers might be framed in the following form: *in the light of the potential costs of control measured against the potential benefits in the form of avoided damage, is further action to reduce air pollution justifiable and necessary?*

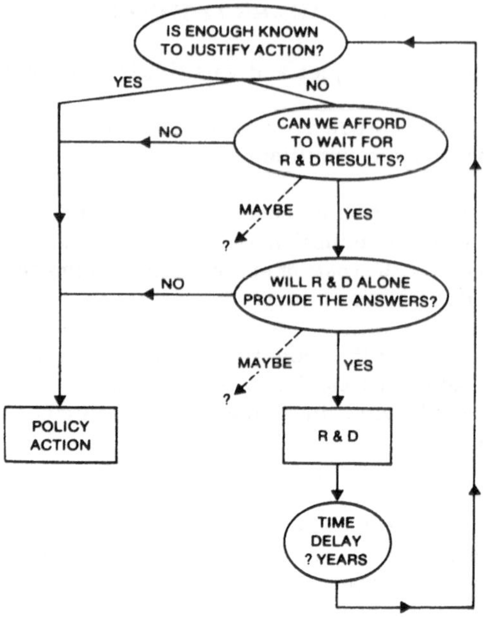

Figure 2: The Policy Dilemma

Some would argue that we do not know enough about the mechanisms of damage to the environment to answer this question in a comprehensive manner. If that is the case, we must ask ourselves: *can we afford to await the results of further research and assessment before taking action, or should action be taken even under a degree of uncertainty,* on the basis that accumulating damage may greatly increase the ultimate costs to society?

In this regard, there is one further question which policymakers need to ask about the timing of a policy response and the extent of that response. *In view of the complexity of the processes involved, can we expect further R&D alone to inform us reliably of the extent to which a reduction in pollutant emissions will be effective in reducing environmental damage?* This is a crucial question in the present debate, because if it cannot be answered in the affirmative, it becomes increasingly difficult, as the evidence of damage caused by air pollution mounts up, to justify further delay in control action on the grounds that further research and assessment are needed. It may then become necessary to complement the R&D with a pollution reduction strategy carefully designed on an international scale. With careful monitoring of the results, these two strategies taken together might provide policymakers with some of the answers and help them to gauge the need for further action in this field.

REFERENCES

Electric Power Research Institute (EPRI), 1983, "Acid Rain Research: A Special Report," EPRI Journal, Vol. 8, No. 9.

Environmental Resources Limited (ERL), 1983, "Acid Rain--A Review of the Phenomenon in the EEC and Europe," published by Graham and Trotman for the Commission of the European Communities, Brussels.

Federal Parliament, 1984, "Third Report on Emission Protection," No. 10/1354, Bundestag, Bonn.

ICF, 1984, "Analysis of the Waxman-Sikorski Sulfur Dioxide Emission Reduction Bill (H.R. 3400)," U.S. Environmental Protection Agency, Washington DC.

Multilateral Conference on the Environment, 1984, "Report of the Causes and Prevention of Damage to Forests, Waters and Buildings by Air Pollution in the Federal Republic of Germany," Federal Ministry of the Interior, Bonn.

OECD, 1984, "Emission Standards for Major Air Pollutants from Energy Facilities in OECD Member Countries," OECD, Paris.

OECD, 1983a, "Costs of Coal Pollution Abatement," OECD, Paris.

OECD, 1983b, "Coal and Environmental Protection: Costs and Costing Methods," OECD, Paris.

OECD, 1981, "The Costs and Benefits of Sulphur Oxide Control," OECD, Paris.

Royal Commission on Environmental Pollution, 1984, Tenth Report, "Tackling Pollution--Experience and Prospects," Her Majesty's Stationery Office, London.

UK House of Commons Environment Committee, 1984, Fourth Report, "Acid Rain," Vol. 1, Her Majesty's Stationery Office, London.

U.S. Department of Commerce, 1981, Review of Businesses, Department of Commerce, Washington DC.

USEPA, 1984, "National Air Pollutant Emission Estimates, 1940-1982," USEPA, Washington DC.

ACID RAIN: OF pHISH AND pHORESTS

The following is a poem written by Dr. Ian Torrens for the conference.

It all began in Southern Sweden
In prior times an angler's Eden.
Authorities in Scandinavia
Blamed it on the Brits' behaviour
Tall chimney stacks to stop the smogs
And distribute them in Scottish bogs
Succeeded in their prime objective,
But led to heightened Scan invective,
Since air pollution blithely coasts
Eastwards over frontier posts
On prevailing winds to roam
Towards the ancient Vikings home
There to fall in rain or snow
Making lakes' pH too low.
Fish populations through acidity
Succumbed with unforeseen rapidity.

The UK, first to feel the chill
of falling Scandinavian goodwill
Was joined by Europe's other nations
Where acid rain had strained relations.
The OECD in '72
Was called upon some work to do,
To investigate the real extent
Of air pollution's touristic bent
A modelling exercise was done
And showed that the phenomenon
Was worthy of increased attention
Hence the need for a Convention.

But chimneys with a lengthy flue
Are frequent in East Europe too
Not members of OECD.
So negotiations went to ECE
A UN body, not to be
Confused with the Brussels EEC.
Upwind and down, both West and East
Came to Geneva, the goal at least
To halt acidifying trends
By cutting pollution that each sends
Across frontiers and on downwind
To fall on sinners and the sinned.
Severe political consternation
Gave birth to long negotiation
A Convention finally to agree
It took effect in '83.

The Nordic countries became the hub
Of the so-called "Thirty Percent Club"

Of countries each of which committed
To reduce the SO emitted
From its national territory
Between 1990 and '93.

Across the wide Atlantic Ocean
Acid rain became a notion
Which led to numerous debates
'Twixt Canada and United States
Regions within the USA
Called upon others to display
Greater restraint on exports of
Acidic Gas on winds above
To give the gentle rain from heaven
A pH less remote from seven.
The international situation
Was helped when governments of each nation
Signed a Memorandum of Intent
To ease bilateral dissent.
But still to a quite large degree
They were unable to agree
To cooperate to an extent
Beyond Research and Development.

Meanwhile in Europe another impact
Came to the fore to interact
With earlier effects detected
And rapidly became reflected
In public and politic alarm
Resulting from important harm
To central European trees
Suffering from severe disease,
The cause of which was quite obscure.
Suspicions fell on air impure.
Scientists who were asked to try
To understand the reason why,
Said, if with chimneys we must deal
Let's not forget the automobile.
Sturm und Drang and consternation
Shook the entire West German nation,
Too hard a choice it was by far
Between their forests and their car.
Happily for the national mood

Technologies beneath the hood
Could be installed, though not with ease,
To stop car emissions hurting trees.
The Government thus legislated:

Emissions must now be abated--
Our chimneys and our cars must be
Much cleaner by 1993.

But Germany is just one part
Of the European Common Mart.
The Brussels Commission began to
broach
The idea of a common approach
Uniting European forces
To reduce pollution from all sources.
Some Draft Directives formulated
Are now intensively debated.
The final outcome's still unclear
But may evolve within a year.

Let's hope these efforts do not fail,
So history can tell the tale
Of rising pH, healthy trees,
And other happy trends like these.
But no silver lining's without a cloud,
And though I should not speak too loud,
When the politics of acid rain
No longer are a cause of strain
I think we all may tend to miss
Travelling to Conferences like this!

Ian M. Torrens
November 1984

ACID RAIN SCIENCE: STATE OF THE ART

SYNOPSIS OF SCIENTIFIC UNDERSTANDING OF ACID DEPOSITION AND ITS EFFECTS

J. Christopher Bernabo*

Executive Director
National Acid Precipitation Assessment Program
722 Jackson Place, N.W.
Washington, DC 20006

Acid deposition research has moved scientists into a complex multidisciplinary arena where new results often appear inconsistent with existing hypotheses. New hypotheses are being proposed and tested, and recent data are showing that some earlier ideas about acid deposition may have been oversimplified, unrealistic, or incorrect. This paper describes some of the recent evolution of our scientific understanding of atmospheric deposition, and the role of that information in developing environmental policies.

NATURAL ACIDITY OF PRECIPITATION

Early estimates of the natural pH of precipitation were based solely on the equilibrium of carbon dioxide in the atmosphere with "pure" water. A pH of 5.6 was subsequently chosen as the baseline against which the seriousness of current acid precipitation levels was judged. It is now clear that other natural factors, such as organic acids, naturally-emitted sulfur and nitrogen compounds and alkaline dust, also affect precipitation's normal acidity. Any baseline pH must account for these factors and the 5.6 value is no longer appropriate.

Using pH 5.6 as a precipitation acidity baseline led to the conclusion that some areas in the eastern U.S. receive precipitation 25 times more acidic than expected in natural precipitation. More recent studies in remote parts of the world on natural sources of atmospheric acidity suggest the unpolluted pH of precipitation is closer to 5.0 than to 5.6. Those regions of the United States experiencing the most acidic annual average precipitation are now estimated to receive 6 to 7 times the average background natural acidity of remote areas.

Estimating the natural pH of precipitation remains a research issue. Comparing the present precipitation pH in the United States with that at remote locations to quantify historical change has limitations. The most notable is that the pH of precipitation is determined by many chemical substances; thus, the measures at

*This paper was presented at the ARIC Conference by Derek Winstanley.

different locations can be the same for quite different reasons. As a result, researchers are now concentrating more on quantifying precipitation chemistry as a whole than on pH alone, focusing on changes of ecological significance from both alkalines and acidic substances. Measuring substances in precipitation like sulfur, nitrogen, and calcium in both polluted and unpolluted environments will allow better estimates of historical change in precipitation chemistry and its significance.

Awareness of natural acidity and total chemistry as they influence the environment has changed our thinking not only in documenting trends in precipitation chemistry but also in the approach to both effects and atmospheric research.

RELATIONSHIP BETWEEN EMISSIONS AND DEPOSITION

When considered globally and over many years, the axiom that what goes up must come down is true as far as emissions are concerned. The uncertainty focuses on when, where, and in what form emissions are deposited on less than global spatial scales and over shorter time periods.

The relationship between emissions and deposition is complicated by atmospheric variability and chemical transformations in the atmosphere during transport. Furthermore, acidic and acidifying substances are found not only in rain or snow, but are also deposited as dry gases and particles. Wet deposition can be accurately documented, but no standardized scientific techniques yet exist to measure dry deposition, which may account for 25 percent or more of the acidity deposited in some areas. The timing, amount, and composition of this wet and dry deposition can influence the effects on the sensitive receptors.

Based on the limited information available, it is now suggested that, when averaged over the entire eastern half of North America for a year or more, there appears to be nearly a one to one (linear) relationship between sulfur dioxide emissions and sulfate deposition. This relationship does not necessarily hold true for smaller spatial scales or shorter time frames. A 50-percent reduction in sulfur dioxide emissions resulting in an overall average 50-percent reduction in wet sulfate deposition over a large area, does not guarantee a 50-percent reduction at a specific site. We are still unable to reliably predict deposition at specific sites or over sensitive regions of critical concern. This complicates designing optimized control strategies or providing strong assurances that a given reduction in emissions will result in a specified environmental benefit.

The acid deposition policy debate has focused primarily on sulfur dioxide. The role of nitrogen oxides, oxidants like ozone or hydrogen peroxide, and other factors interrelated with acid deposition is uncertain, but atmospheric and effects studies are making it clear that these other factors are also important.

Sulfates contribute about two-thirds of the acidity to precipitation in the Northeast; but, nitrates can contribute over half of the acidity in western regions of the United States. In addition, most of the acid deposition policy debate has focused on long-distance transport, even though local sources are believed to be significant in many cases. The relative role of local versus distant sources varies greatly among different receptor sites and remains a major uncertainty in evaluating the effectiveness of possible control strategies. The National Program is examining local as well as distant inputs to deposition.

Improved understanding of the complex processes of transport, transformation, and deposition is essential for the reliable assessment of the consequences of various control strategies. The route to improved understanding of these processes is to conduct research to test existing hypotheses and to develop improved tools for reliably predicting the relationships between emissions and deposition.

A number of research approaches are being pursued by the National Acid Precipitation Assessment Program (NAPAP), and others, the main ones being laboratory studies, field studies, and mathematical modeling. Laboratory studies are improving our understanding of pollutants emitted into the atmosphere, of chemical reaction rates and mechanisms, and of the products of these reactions.

Mathematical simulation models provide a basis for integrating and interpreting the research studies on atmospheric transport, chemical and physical transformation and removal, and for quantifying source-receptor relationships. Existing models have fundamental weaknesses in that a number of phenomena have not been treated fully and the chosen parameters of the various processes have not been coupled with real-world validations. With advances in scientific research and in computer technology the field is ripe for progress.

To meet the assessment objectives of the National Program, an advanced Eulerian Regional Acid Deposition Model (RADM) is being developed. The RADM is being assembled in modular form from existing modeling components and will incorporate more realistic definitions of processes as they become available. To aid the interpretation of modeling results and the proper assessment of source-receptor relationships, efforts are focusing on statistical and uncertainty analysis. As they become available, the results of the Eulerian modeling efforts will feed into the assessment activities of the National Program. Canada is also supporting the development of an Eulerian model, and scientists in both countries are working cooperatively.

The development of other less complex models is also contributing to our improved assessment capabilities; these include models that can be used to assess the effects of nearby point or urban sources of acidic materials and precursors, and models to provide complete sulfur and nitrogen budgets for North America. Model intercomparisons are also in progress.

Theoretical models by themselves, however, cannot provide the necessary scientific credibility in determining source-receptor relationships and must be evaluated by comparing model results with actual observations. Therefore, large-scale field studies are needed to provide the data to confirm model assumptions and model characteristics. In the past, the lack of model evaluation has been a key factor in the scientific uncertainty attached to numerical models and, hence, in their limited use for assessment and policy analysis purposes.

An alternative method to numerical models for determining source-receptor relationships is empirical determination. Field experiments for this purpose will, of necessity, be large and costly; feasibility studies are now being undertaken. Technology is also being refined that will allow the improved tracking of pollutants in the atmosphere and their deposition on the earth's surface.

To insure that the various field experiments are planned and executed in a coordinated and cost-effective way, a Working Group on Atmospheric Field Experiments has been established that involves scientists working under the National Program, in the private sector, and in Canada.

AQUATIC EFFECTS

Predicting biological response to acidification depends not just on measuring acidity itself, but also upon knowledge of concentrations of substances like sulfate, calcium, aluminum, and organic matter. Even when acidity (pH) is equivalent between two similar lakes in the same region, observed fish populations can differ widely. Different concentrations and the interactions among the numerous substances present in the water, such as aluminum and organic matter, may be the key. Scientists believe that fish do not successfully reproduce in some clear acidic waters largely because toxic aluminum concentrations may occur as acidity increases. Such effects work slowly and rarely result in massive kills of total fish populations.

Aquatic chemists are still unable to predict reliably the rate of change in surface water chemistry that is caused by a certain rate of acid deposition. Some surface waters have been affected by acid deposition, but how many still must be determined. Hypotheses based on currently available information predict rates of surface water acidification ranging from no significant changes at present levels of acid deposition to predictions that numerous systems will become increasingly acidic in the near future. An increased awareness of the influence of soils and vegetation on surface water chemistry is largely responsible for this divergent thinking. It is not yet possible to determine which predictions are correct and under what conditions.

Declines in pH and alkalinity, a measure of acid-neutralizing capacity, were first described in Scandinavian lakes characterized as nutrient poor, surrounded by thin (or no) soils, underlain by granite, with steep slopes. From the chemical changes observed, it appeared that in such lakes acid deposition consumed the acid-neutralizing capacity of the watershed and the associated water bodies to a point at which acidification could occur. In systems with such limited acid neutralization capacity for sufficiently acid soils it is believed that once the neutralization capacity is reduced, surface water quality will resemble precipitation chemistry within years to decades. Surface waters with these response characteristics have been identified in a few areas in the United States, and represent one category of waters and watersheds commonly referred to as "quick response" systems (changing in time frames of years to decades).

Quick response systems represent the most sensitive class of surface waters. In contrast to these systems are "no response" systems, those subject to change in time frames of centuries or millenia. Wet and dry deposition of acid and acidifying substances typically pass through the vegetation canopy and over or slowly through soils before entering water bodies. At any point along this pathway, various processes can neutralize the acidity of the precipitation. Many terrestrial systems offer sufficient neutralizing capacity to prevent the bulk of most highly acidified precipitation from entering a water body in an acidic form. In addition, runoff from agricultural activities may be expected to overwhelm natural inputs.

Between the two extreme types of surface water systems are those referred to as "delayed response" systems, those subject to change in the time frames of decades to centuries. Four major processes control the rate at which these systems might be acidified:

- Acidity and amount of deposition.
- The release of base cations to the water flowing through the soil.
- Retention of sulfate by the watershed.
- Biological transformations within the terrestrial and aquatic systems.

Assuming that deposition of acidic and acidifying substances remains the same, the latter three processes control the rate of acidification, and collectively could be referred to as neutralization buffering capacity. The rate at which neutralization capacity is chemically produced or consumed in the soil is controlled by the release of cations like calcium and retention of anions like sulfate; soils differ widely in their neutralization capacity. A myriad of other processes, both chemical and biological, including the hydrologic route and rate of water movement through soils, may be important. Delayed response systems, found in all regions of the United States, are subject to acidification when sulfate retention has been exhausted and base cation release is less than the amount of acidity being introduced by acid deposition.

The amount of sulfate that can be retained (adsorbed) in a soil depends on the chemical composition of soil, temperature, vegetation type and growth, and soil depth. Surface water acidification can occur only after the soil's capacity to retain sulfate or other acid anions like nitrate is exceeded and these substances are able to leave the soil and enter lakes and streams.

Movement of acid anions like sulfate and nitrate from the soil to lakes and streams does not necessarily result in surface water acidification. Acid cations like hydrogen or aluminum must move with the sulfate anion for acidification to occur. If sufficient base cations such as calcium and magnesium are present, they move with the acid anions, therefore contributing little if any acidity to surface waters.

In general, there are two pools of base (acid neutralizing) cations in soils:

- Those on soil exchange sites which are adsorbed on soil particles and are readily reactive.
- Those that form soil minerals derived from rock, and biomass which represent a large, slowly reactive reserve.

Once the input of acids exceeds the rate of release of base cations from either pool in the soil, precipitation is not completely neutralized and it proceeds to surface waters in acidic form.

In quick response systems, where sulfate retention is small and sufficiently acid soils exist, acidification usually occurs rapidly as a result of acid deposition. Lakes and streams associated with watersheds having a high ability to retain sulfate but low ability to supply base cations will acidify at slower rates. Those watersheds with a high ability to supply base cations are protected from change. The major question is the extent and location of these different type systems in the United States.

For delayed response systems, there is uncertainty concerning the rate at which soil and surface waters will respond to acid input. For example, in systems where sulfate adsorption capacity is high and cation release is low, it is not certain that associated surface waters would be acidified. Theoretically, these soils could deteriorate in their ability to assimilate acid inputs. As acid inputs continue, less and less of the deposited sulfate would be retained, and the surface waters could become increasingly acidified. However, where sulfate adsorption capacity has been exceeded and associated surface waters are not now acidic, it is unclear whether these systems will acidify in years to decades. In these systems, cation release may be sufficiently rapid to renew this portion of the soil's neutralization capacity, thereby preventing further deterioration of the surface water. In these instances, there is disagreement as to whether further acidification will be gradual over periods of centuries or perceptible in years to decades. The disagreement stems from a lack of data.

To resolve these debates, we must determine the rates of cation exchange and renewal (mineral weathering rates) and how they are affected by acid sulfate adsorbing soils, and quantify the sulfate retention capacity of soils. Other variables, such as the hydraulic pathway of water through terrestrial systems, which determines the contact time between soil and precipitation and thus contributes to the potential for neutralization, must also be examined as factors contributing to the acid-neutralizing capacity of ecosystems.

All these issues are being investigated through the National Acid Precipitation Assessment Program. Making a valid assessment of how many aquatic systems will become acidic in the future depends on the results of research now in progress to define and quantify the processes involved. The assessment also requires an accurate inventory of terrestrial and aquatic ecosystems categorized by characteristics determined to be indicators of potential acidification.

TERRESTRIAL EFFECTS

Soil scientists generally believe that most managed agricultural soils are not very vulnerable with respect to acid deposition. Materials routinely added to such soils in the form of fertilizer and lime overwhelm the influence of atmospheric deposition. For unmanaged soils, recent studies on natural soil acidity production have decreased concerns over soil pH changes. The deposition of acidic substances on soils may be making only a minor contribution to the natural soil acidification processes. As noted in the previous subsection, it is now believed that understanding the previously overlooked sulfate adsorption and mineral weathering rates in soils is crucial to understanding how soils react to acid deposition. Both processes vary widely in ecosystems of North America.

The sulfur and nitrogen in acid precipitation are primary plant nutrients. Because of this, acid deposition may have both positive and negative effects on plant growth. Scientists have found little conclusive evidence for major direct effects of acid precipitation on vegetation at current deposition levels. The direct effects on crops appear less significant than those related to other air pollutants such as ozone. In studies where specific crop varieties have demonstrated sensitivity, other varieties or cultivars of the same crop were found not to be sensitive. Although future research is required to increase the certainty of our conclusions, concern over crop losses has decreased substantially.

For forests, the story is different. Initial studies and hypotheses on forest tree productivity indicated that forests were not expected to be threatened by acid deposition. The primary constituents of acid deposition, nitrogen and sulfur, are essential nutrients for optimal forest growth. Nitrogen deposition was even expected to increase production in most cases because inadequate nitrogen can limit forest growth. Indeed, in some parts of Scandinavia, forest growth apparently has increased. The balance of sulfur, on the other hand, rarely limits growth in forest ecosystems; it is believed that direct damage caused by excess sulfur is rare. However, observed reductions in tree-ring widths and forest dieback in areas with elevated pollution levels and high acid deposition, such as the eastern United States and central Europe, have heightened concern about the potential negative effects of acid deposition and other air pollutants on some forests.

Within the past few years, West German scientists have observed increasing evidence of forest damage. The forest area showing damage is expanding rapidly. A survey in 1980 indicated that 8 percent of West German forests had some trees displaying evidence of damage. The 1983 survey indicated that damage was evident in 34 percent of the forest regions, and the 1984 estimate is 55 percent. This may represent an environmental change of unprecedented rapidity.

The specific causes of these observed forest changes are still being debated, but air pollutants are prime suspects. While the initial concern was about sulfur dioxide and acid deposition, many scientists currently believe nitrogen oxides and oxidants such as ozone may be contributing to the observed forest growth changes. Several causes may be acting together and many different hypotheses have been proposed in West Germany to explain the observed forest changes. Most of the affected forests in West Germany are closer to emissions sources than such areas in the United States and have roughly double the ambient levels of polluting gases while receiving precipitation of similar acidity.

Under the National Acid Precipitation Assessment Program, scientists are beginning to look more closely at forest growth history in the United States. From initial surveys of forests in the eastern United States it is becoming evident that the symptoms of decline are somewhat different than those being observed in West Germany, but a decline in tree-ring widths over the last 20 years is measurable in a number of eastern forests. The slowing radial growth indicated by tree-ring analysis suggests the possibility of a broad regional phenomenon observed in several species and in both young and old trees. In some high elevation areas, visible dieback of certain coniferous tree species also has occurred and acid deposition is being investigated as one potential cause. Several hypotheses are currently proposed to explain the dieback and declines in tree growth both in West Germany and the United States.

1. Gaseous air pollutants, such as ozones, are causing damage to tree foliage and, either alone or by interacting with acid deposition, are affecting tree growth.

2. Deposition of metals, alone or interactively with acid deposition, is directly or indirectly affecting tree growth.

3. Acid deposition is increasing the leaching of nutrients from the foliage of trees at a rate sufficiently rapid that uptake and recycling of lost nutrients cannot maintain adequate supplies of nutrients in the tree foliage.

4. Acid deposition is acidifying soil water, increasing aluminum concentrations, and ultimately affecting root growth and nutrient and water uptake.

5. Nitrogen saturation of forest soils is changing the beneficial relationships between trees and soil microorganisms either directly or indirectly, resulting in nutrient and water imbalance or reduced winter hardiness.

All of these working hypotheses appear to have some merit. Indeed, many of the factors proposed to have contributed to forest decline may be occurring simultaneously because they are not independent of each other. For example, increases in ozone, metals deposition, and acid deposition all appear to have occurred concurrently in the eastern United States. Further research is needed to determine which factors primarily cause the observed dieback and/or declines in growth. Because consideration must also be given to natural environmental stresses such as drought and pests and potential interactions with acid deposition, the issue becomes exceedingly complex.

The observed reduction in tree-ring widths in the United States began in the late 1950's and early 1960's. Climatic fluctuations alone do not presently seem to fully explain these declines. Man-made causes are thus prime suspects. Research is underway to investigate the role of acid deposition, if any, in the observed growth changes. Among air pollutants it is unclear, however, which are of more concern. Hypotheses suggesting nitrogen oxides and ozone as the primary factors influencing forests have been proposed, while in the case of surface water impacts, sulfur is still believed to be the prime factor of interest.

Intensive study in the United States, West Germany, and elsewhere is now testing the many competing hypotheses. The potential for air-pollutant-related damage to forests from air pollutants is a rapidly growing concern because the economic and social consequences of forest damage could far exceed those related to surface waters. The National Program has greatly accelerated research in this area and has initiated cooperative efforts with West Germany and Canada on this important subject.

MATERIALS AND OTHER EFFECTS

Damage to materials from atmospheric deposition takes a variety of forms including the corrosiön of metals, erosion and discoloration of paints, and deterioration of building stone. All of these effects occur to a significant degree as a result of natural environmental conditions. Moisture, carbon dioxide, sunlight, atmospheric oxygen, temperature fluctuations, and the action of microorganisms all contribute to the deterioration of materials. Quantifying the specific contributions of anthropogenic air pollutants to such damage is a complex task. Furthermore, distinguishing the relative amount of damage caused by specific pollutant transformation and contact processes (for example, acid precipitation) becomes even more difficult.

The percent of materials degradation that occurs from acid deposition, in contrast to the numerous other natural variables which cause degradation, is not known. The distinction between the effects of dry (including gaseous and particulate) and wet deposited substances is difficult to assess. The primary factor in degradation of materials due to acid deposition is probably the corrosive action of the acids themselves. These acids are formed from sulfur and nitrogen oxide transformation. It appears that the impact of dry deposition on moist surfaces or dry deposited materials in a very humid environment is important for understanding the effects on materials. Under these conditions, the acids can become highly concentrated doing excessive damage.

Determining the direct effects of acid deposition on materials depends on our understanding of atmospheric corrosion, and the chemical processes of materials weathering and deterioration. Our understanding is directly related to our ability to discern through a coordinated set of laboratory and field exposure studies how sulfuric and other atmospheric acids attack materials; this includes not only direct surface corrosion, but secondary chemically induced, stress-related failure resulting from weakened mechanical strength.

The direct effects on human health of inhaling sulfate particles and aerosols continue to be studied under other pre-existing federal programs. Studies in the National Acid Precipitation Assessment Program of acid deposition's relationship to human health focus on the new area of indirect effects after these substances are deposited. In this context, no significant indirect human health effects have yet been documented in the United States as being a consequence of the deposition of acid. The potential health impacts related to drinking water and the food chain are covered under the aquatic and terrestrial effects portions of the National Acid Precipitation Assessment Program.

The National Program's first survey of New England municipal drinking water supplies did not indicate any demonstrable effects from acid deposition. However, data on the impacts of acid deposition on drinking water quality are still scarce, with the most serious lack of data in the groundwater effects area. Increasingly corrosivity of acidic water is probably the most significant potential impact. Populations could be exposed to higher concentrations of corrosive toxicants if water becomes more acidic.

The complexity of the acid deposition phenomenon and its effects is such that definitive answers to many key questions are not yet available. Scientific thinking about acid deposition issues has changed and is likely to continue to change. Fortunately, given the acceleration of research under the National Program, by the private sector, states, and other nations, major strides forward in our understanding are expected in the coming years.

CONCLUSION

The research and monitoring activities of the NAPAP Program are providing successively better tools and a continually improved information base for developing and implementing sound policies to achieve the maximum environmental benefit at minimum cost to society. As Congress recognized in creating this long-term, broad-based program, the issue is exceedingly complex and has required advancing the frontiers of science in fundamental areas of ecology and atmospheric sciences.

Addressing the acid rain issue effectively requires both research and appropriate environmental policies. Objective scientific information is only one contribution, albeit an important one, to developing environmental policies. Policymakers, not researchers, must decide when scientific information is adequate for decisionmaking. The scientists can define at any point in time what is known, with what level of certainty, and what is not yet known.

Policymakers must also make a host of subjective judgments for which scientists have no special expertise. For example, scientists have the task of relating the response of ecosystems to the amount of acid deposition they receive; but it is the role of the policymaker to determine whether emissions should be limited, and by how much, considering the social costs and benefits of prescribed actions.

Although one major goal of the National Program is to quantify such costs and benefits, often they cannot be quantified reliably, and subjective value judgments must supplement scientific analyses. Science does not provide a basis for making such value judgments; those judgments are the role of the policymaker. However, enhanced scientific understanding can strengthen our confidence in selecting the most effective actions to yield the desired environmental results.

DISCUSSION

MYRON UMAN (National Academy of Sciences): The NAPAP presentation was excellent in every regard. I would like to make just a couple of comments. First--as to what the natural value of the pH rain might be. This is a highly variable function which depends upon biological as well as geological conditions, locally or regionally. As has been pointed out--and I'd like to emphasize again--the value which is frequently quoted as the natural pH of rain of 5.6 is a calculated value based on equilibrium between atmospheric concentrations of carbon dioxide and distilled water. In no case in the real world do we have that. We have acidic substances in the air in addition to carbonic acid, as well as alkaline substances, so depending on where you are the natural value of pH can range below or above 5.6 but in most of the regions of the world where there is a lot of vegetation and not a lot of wind-blown dust we would expect the natural pH to be less than 5.6.

I would like also to make a comment about the paper mill near Hubbard Brook. For some reason, I guess there has been a lot of excitement about this. We also have month-to-month emissions data from this facility from the mid-1960's to 1977, for both sulfur and chlorine. Sometimes this plant was closed down for months at a time. We also have the Hubbard Brook deposition data on a week-to-week basis for a variety of substances including chlorine. A detailed statistical analysis using the temporal variations suggest that it is probably not possible that the plant had more than a 1% influence on the deposition data at Hubbard Brook. The strength of this argument is based on the chlorine record.

One more thing I would like to point out, and that is that as far as the processes of acidification of lakes is concerned, it is possible to resolve natural acidification processes from short-term influences of man. That is, it's possible to look at lakes--particularly in the sedimentary record of lakes--and infer long-term changes in the water column pH. Lakes have been undergoing acidification naturally since the glaciers retreated but that process is a very long and slow one. There are data, which we expect to publish in March, showing a very slow acidification process over the last century to a century-and-a-half in particular lakes, and on top of that a very substantial increase for rates of acidification over the last decade or two.

In talking about lake acidification data, however, and particularly about the surface water chemistry data of lakes, it is important to remember that the lakes that are best suited for assessing the effect of acid deposition itself are very few, and that they are atypical of the lakes in any given region. They are atypical in the following sense: they not only have to be geologically sensitive but they also have to be lakes that have been unaffected by man in other ways, which means that the land-use practices in a watershed have to be essentially nonexistent. There are few lakes where you can really say this, and one of the difficulties of large-scale lake surveys is that it's rare to have land-use data for the lakes. In those circumstances where land-use data are acquired at the same time, perhaps we can determine whether what we are looking at is the result of acid deposition or a logging practice or agricultural practice or some other kind of activity. Lake data need to be stratified. For example, in certain parts of the country, acid-mine drainage may be much more important to water bodies than acid rain could ever be. Nonetheless, acid rain may still be occurring in these regions, but the influence of mine drainage is too strong to notice that of acid rain. Thank you.

MIKE RUBY (Seattle, Washington): You have approached your presentation with the implicit statement that the acid nature of the deposition is what we are interested in, in other words, the hydrogen ion content. I was under the impression that it was much more uncertain than that--that we really weren't certain whether it was the anion or the cation that might be at work in the effects, particularly in terms of the forest effects and that, in fact, in terms of dry deposition the SO_2, the actual sulfur input, might be as important as the sulfate or the hydrogen ion accumulation. I don't know any of this and I am hoping that you or someone else here, a biologist or someone of that persuasion could enlighten us about the current thinking on this.

DEREK WINSTANLEY: Certainly on dry deposition it may well be important, we just don't really have enough information on dry deposition to be able to make any sort of definitive statements about it. As to the relationship of the anion and cations, I think that I pointed out that, in the watershed study, one of the key results was that it is extremely important to look at the total ion content of watersheds. I think that is now well recognized in the research and we are not just restricting ourselves to the hydrogen ion, but taking a broader look at total ionic balances. We are also looking at neutralizing substances in the atmosphere that can neutralize some of the acids.

MIKE RUBY: I'm sorry it must have been not understood, I was speaking in terms of the effects, not what the researchers were collecting. Is it believed that the effects are directly due to pH and hydrogen ion content, or is it believed that the effects might be effects derived from sulfate as sulfate, or nitrate as nitrate?

DEREK WINSTANLEY: I can't answer that question; I am not sufficiently informed about those aspects of effects research. I suggest that you talk to some of the effects researchers.

RAY EFFER (Ontario Hydro): I would like your comment on this: I think it has to be emphasized that we are spending a lot of time measuring what we can do easily and what we can't do easily but, we are tending to ignore dry deposition and by that I mean mostly gaseous deposition. It seems to me that we are making a great fuss over the nonlinearity question and that is predicated primarily on whether we're depositing sulfate or sulfur. A fish doesn't care whether it is hit on the nose with a sulfur dioxide molecule or a sulfate molecule or a drop of sulfuric acid, and what we have to do is to emphasize total sulfur, total acidifying potential, rather than emphasizing what we can measure so easily--which is wet deposition.

DEREK WINSTANLEY: I agree entirely, and this is why the national research program is giving great emphasis to developing techniques that allow us to measure dry deposition and to establish a dry deposition network. We would recognize that it is equally as important to understand what is happening in dry deposition, both in terms of monitoring it and understanding the effects of dry deposition.

RAY EFFER: I think it has a direct influence on the purpose of this conference that we can produce deposition isopleths which might influence economic thinking, but how that acidity reaches the ground as a wet or a dry form may produce isopleths which are totally different, and I might suggest, would require quite different source control mechanisms.

DEREK WINSTANLEY: I agree entirely. I stressed through my talk that we can not measure dry deposition very well and that it is probably equally as important as wet deposition.

GEORGE TOMLINSON (Domtar in Montreal): I'd like to point out that I think dry deposition is probably more important than wet deposition in many locations. At Woods Lake it was found that spruce were catching, I think, about 118 kilograms of sulfate per hectare as against around 40 from wet deposition so it is really 3 times as much. Studies at Dartmouth on Mt. Lusilocki and the University of Vermont showed values of around 250 kilograms per hectare of sulfate deposited on the tops of the mountains because of dry deposition and interception deposition of acidic cloud water so that when we draw a certain map, such as you showed, showing the amount of sulfate deposited we completely ignore these "hot spots" and these are the "hot spots," or I should say the major locations, where the tree damage is most serious, and what is happening is that sulfate ion is extracting the nutrients--calcium and magnesium--from the soil and the trees are actually showing calcium and magnesium deficiencies.

In Germany where I have been following the case very closely, in 1980 there were a few signs on the mountains; everybody said, "Forget it; they're not commercial forests; those things are subject to odd weather," and so on, and what's happened is that the problem has moved down. In Holland, they believe that practically all their trees are affected. One of the factors there is ammonia, which actually liberates large amounts of the hydrogen ion when it's involved in the soil, and has quite the reverse effect of its effect in the rain. I think that we need to have much more careful study of this from the standpoint of the forest. What is missing from our networks in North America is the study of SO_2 and sulfate concentrations, which they have been doing in Germany. It shows that very low concentrations can result in very high dry and wet deposition, particularly on spruce, which has very large needle area--surface area--for deposition uptake. So I would like to urge you to at least establish SO_2 and sulfate stations in rural areas.

DEREK WINSTANLEY: We are doing this. This year we are starting to establish what we call a cloud-monitoring network at high elevation stations, starting in the southern Piedmont and going up through the Adirondacks, monitoring exactly what is being deposited in those high elevations.

JAY JACOBSON (Boyce-Thompson Institute): I also enjoyed your talk very much and thought it was an excellent summary of what we have learned. I'd like to add a comment which relates to some of the questions that have been raised in regard to your talk. Research on effects of acidic deposition on ecosystems, terrestrial and aquatic ecosystems, in the U.S. has undergone very marked change in the last decade. And that shift has been from emphasis on acidification--the hydrogen ion--to an emphasis on the other components of rain along with the hydrogen ion, and their influence on ecosystems. We have learned, for example, as Dr. Tomlinson just pointed out, that sulfate has very, very important effects on ecosystems, but its effect varies greatly with the nature and characteristics of the individual ecosystem. We are also beginning to learn, as you pointed out, that nitrogen deposition has important effects on ecosystems, but it's quite likely that there too its effect, the effect of nitrogen, varies greatly depending upon the characteristics of the ecosystem. So the emphasis now is, in fact, on all of the components of acidic deposition, not just on hydrogen.

DAVID LANTZ (Hoosiers for Economic Development): I guess I am one of those people that is interested in the paper mill at Hubbard Brook. My question is, first of all, was this discovered more or less by accident, or was the paper pulp mill discovered through intense research effort? Secondly, if it was discovered by accident, is anyone now looking to see if these and other similar types of establishments exist elsewhere? I realize that one place that has the effect of 1 or 2% may not be much, but perhaps if there are a number of these they could add up. Third, if no one is making a concentrated research effort to see if other establishments do exist, do you think it would be a good idea to make that type of research effort?

DEREK WINSTANLEY: Pete Coffey of the New York Power Pool released an article a few months ago pointing out the existence of this mill. I am not sure whether it was discovered by accident or not. A lot of people have looked at the data very closely since the Academy released its report in 1983 and this is one of the things that people have now been picking up. As far as whether anyone is looking at other stations to see if there are any other significant local influences--I presume they are since this sort of revelation tends to increase people's awareness of these sorts of problems. But I would also like to point out that there are very few stations, if any, with the type of record that Hubbard Brook has in terms of a continuous record over 20 years. So, there may be other local influences on other stations, but as far as interpreting those data, to generalize about relationships between emissions on a regional scale and deposition, Hubbard Brook is probably unique. I agree that this is the sort of effort that needs to be undertaken.

MYRON UMAN: Just a quick comment on the paper mill. It is my understanding that the people who were doing the long-term research on forest ecosystems at Hubbard Brook were well aware of the existence of that mill. It's a very small facility, had a very low smoke stack, located roughly 8 kilometers to the northeast on the other side of a high ridge. Micrometeorology at a place such as this would suggest a very, very small chance of an influence, so they were aware that it was there, but looking at their data and understanding what they were trying to do, they did not assume that there was an influence from the facility on their record.

CONFRONTING THE ASSUMPTIONS AND UNCERTAINTIES: ASSESSING COSTS

COST AND COAL MARKET EFFECTS OF ALTERNATIVE APPROACHES FOR REDUCING ELECTRIC UTILITY SULFUR DIOXIDE EMISSIONS

C. Hoff Stauffer, Jr.

President
ICF Incorporated
1850 K Street, N.W., Suite 95
Washington, DC 20006

INTRODUCTION

It appears the capability to assess the costs and coal production effects of alternative approaches for reducing electric utility sulfur dioxide emissions exceeds the capability to assess the benefits of such reductions. This is unfortunate because any acid rain mitigation program should be designed such that the marginal benefits approximate the marginal costs. Our capability to assess the cost and coal production effects may result in too much emphasis on these measureable effects and not enough emphasis on the value of the environmental benefits being sought.

This is not to say that there are not opportunities to improve the estimates of cost and coal production effects. There are. Two areas that warrant refinement are (a) the effects of higher electricity rates on electricity consumption and (b) the treatment of transportation rates in estimating consumer costs.

The reason that different studies come up with different estimates of the cost and coal production effects of reducing electric utility sulfur dioxide emissions is that different studies:

(1) make different assumptions, and

(2) (sometimes) use different measures of costs.

REQUIRED EMISSION REDUCTIONS

The principal reason for different cost estimates is that different levels of required emission reductions are assumed.

Greater emission reductions result in increasingly higher costs. Total costs almost double for every additional 2 million tons reduced (over six million tons).

Hence, it is not surprising that cost estimates of a 12-million ton reduction greatly exceed those for a 10-million ton reduction or that cost estimates of a 10-million ton reduction greatly exceed those for an 8-million ton reduction.

To maximize society's welfare, the marginal cost of a program should approximate the marginal benefits. The marginal costs accelerate rapidly from about $300 per ton at a six-million ton reduction to $500 at eight million tons to $900 at ten million to $1400 at twelve million tons.

Program designers should exercise good judgment in setting the required reductions. What would one have to believe for the last ton of sulfur dioxide reduced to be worth $1400?

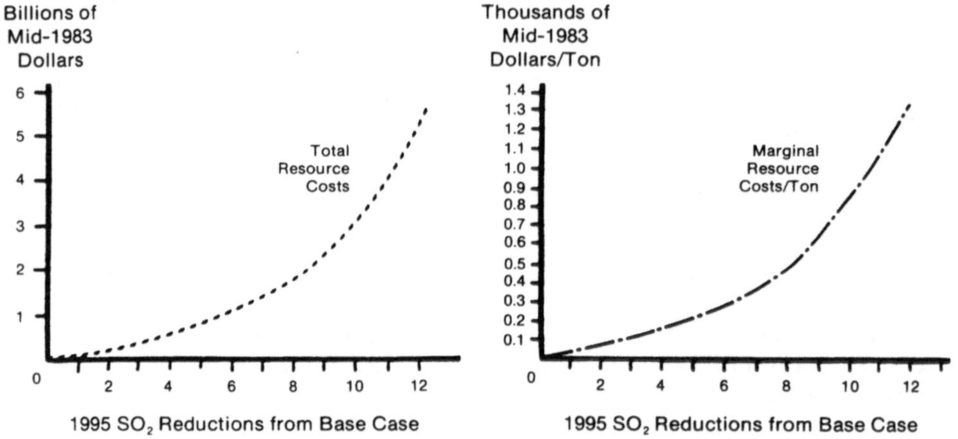

Figure 1. Cost of Emission Reductions

SCRUBBERS

Retrofit scrubbers are generally not cost-effective. Shifting to lower sulfur coal is generally the least-cost option for reducing emissions, until the required emission reductions approach 12 million tons.

Hence, programs that require scrubbers will be more expensive than those that don't.

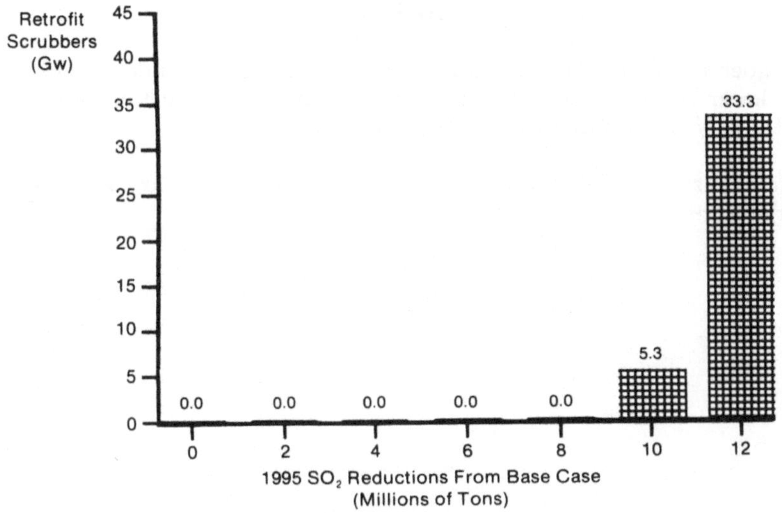

Figure 2. Retrofit Scrubbers and Cost-Effectiveness

SCRUBBER COSTS

Analyses that assume scrubbers will be more expensive will generate higher cost estimates.

Analyses that employ initial-year costs will generate higher estimates of scrubber costs than those using levelized costs, because under standard rate-making practices the capital charges associated with a scrubber are high initially and decrease over the life of the scrubber.

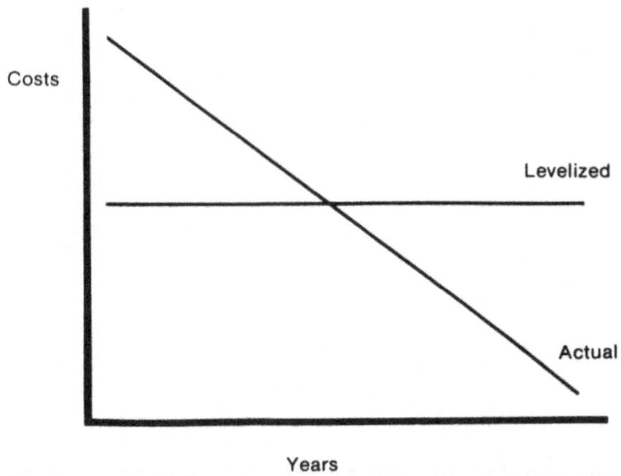

Figure 3. Capital Charges Associated with a Scrubber

COAL CONSUMPTION

Higher levels of required sulfur dioxide reductions would result in greater shifts to lower sulfur coals, until about a 10-million ton reduction when scrubbers become the only viable option for additional reductions.

Hence, estimates of shifts in coal consumption should vary with the required level of reduction.

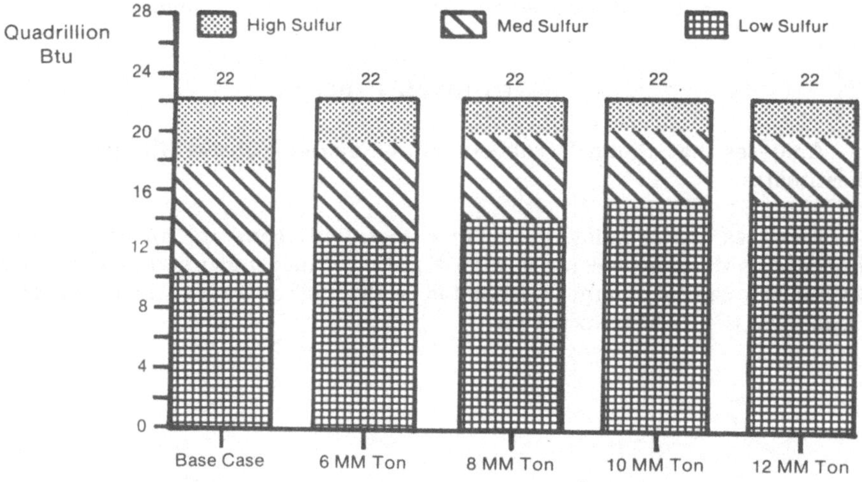

Figure 4. Coal Consumption and Sulfur Content

Some analyses employ an extremely rudimentary approach for estimating shifts in coal consumption. If a higher price differential is assumed, the estimated costs of switching to lower sulfur coal will be greater.

COAL PRODUCTION

Coal production shifts result from coal consumption shifts.

If utilities would shift from higher sulfur coals to lower sulfur coals, production would decrease in higher sulfur regions (e.g., Northern Appalachia and the Midwest) and increase in lower sulfur regions (e.g., Central Appalachia and the Rockies).

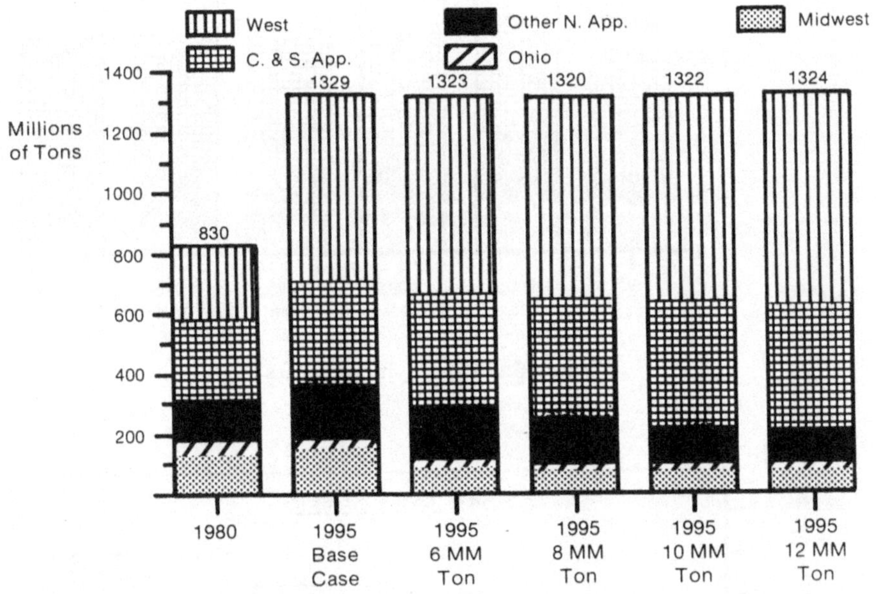

Figure 5. Shifts in Coal Production

EMPLOYMENT

Shifts in production result in shifts in employment. Shifts to production of lower sulfur fuels would shift jobs but not eliminate them. But, employment in high sulfur regions would decrease.

Hence, programs have been proposed that require scrubbers rather than fuel switching.

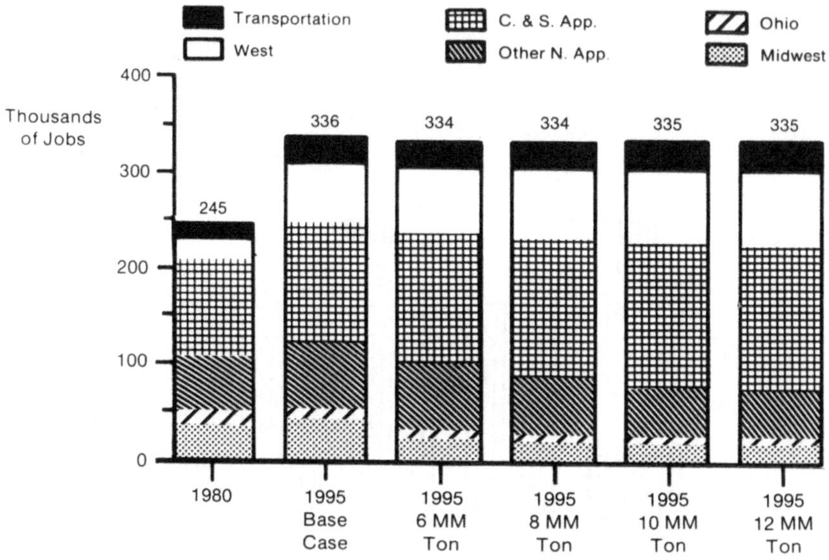

Figure 6a. Employment Shifts

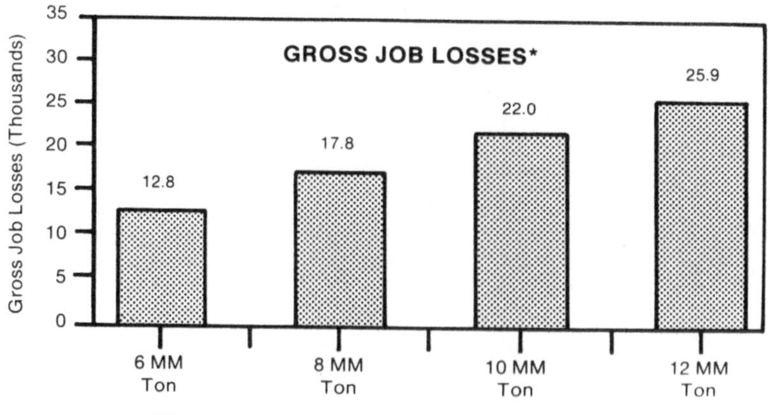

Figure 6b. Fuel Switching and Job Losses

WAXMAN-SIKORSKI

The Waxman-Sikorski bill represents the scrubber approach.

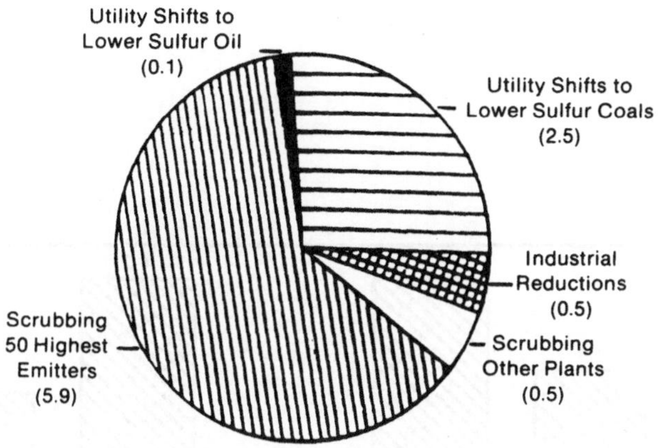

Figure 7a. Waxman-Sikorski Reductions, 1995
Total Reductions - 9.5 Million Tons

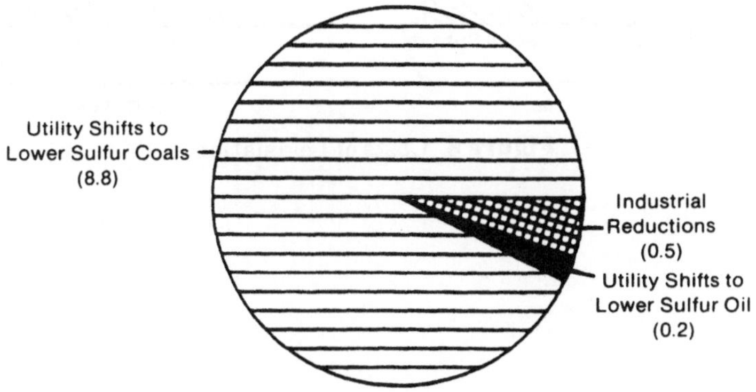

Figure 7b. Cost-Effective Equivalent Reductions, 1995
Total Reductions - 9.5 Million Tons

COSTS

The costs of the Waxman-Sikorski bill exceed those of the "cost-effective equivalent" by about $1 billion per year.

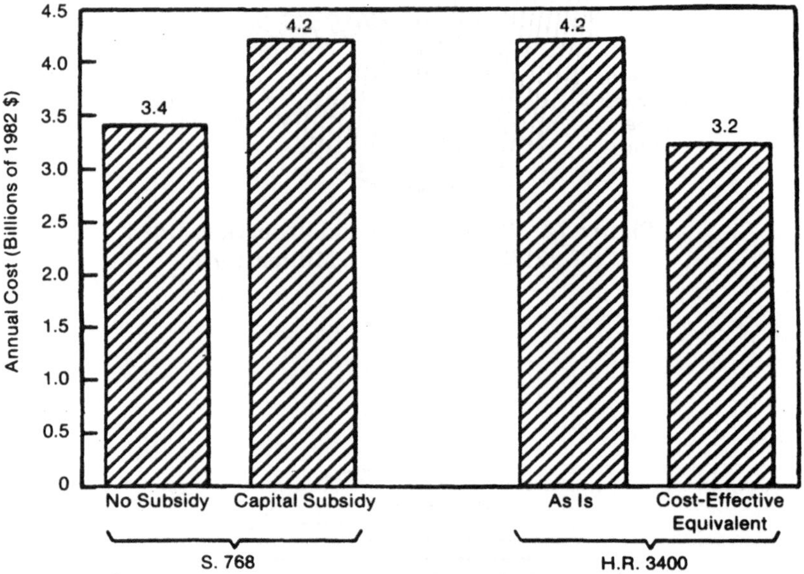

Figure 8. Costs of Legislation

GEOGRAPHICAL DISTRIBUTION OF COSTS

Waxman-Sikorski spreads the cost impacts across states more evenly, primarily by increasing the costs in the lower cost states.

Table 1

Cost Impacts by State

	Present Value of Utility and Tax Costs (billions of 1982 $)	
	Waxman-Sikorski Bill	Cost-Effective Equivalent
Pennsylvania-West Virginia	4.5	2.9
Georgia-Florida	2.1	1.1
Ohio	3.9	4.2
Michigan-Wisconsin	1.8	1.1
Indiana	2.2	2.5
Illinois-Missouri	3.3	2.0
Kentucky-Tennessee	3.2	1.9
Total 12 States	21.0	15.7
Other 19 States	4.6	0.7
Total 31 Eastern States	25.6	16.4
Total 17 Western States	6.6	0.9
Total U.S.	32.2	17.3

COAL PRODUCTION

But, Waxman-Sikorski would prevent coal production shifts.

EMPLOYMENT SHIFTS

Hence, the employment shifts would not occur.

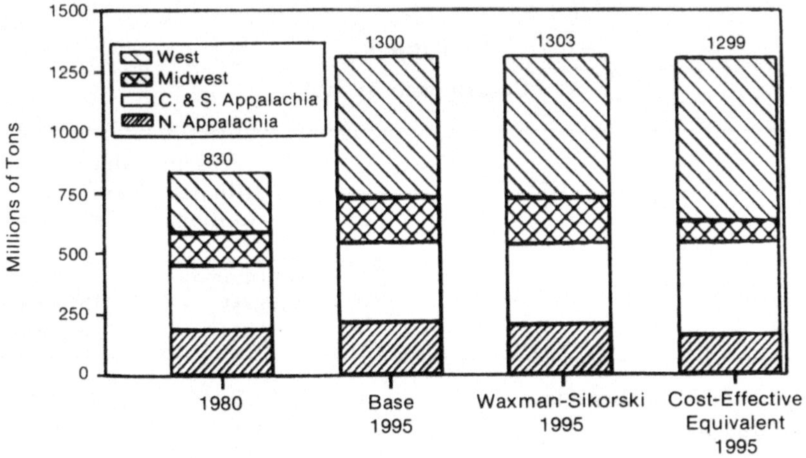

Figure 9a. Regional Production Impacts

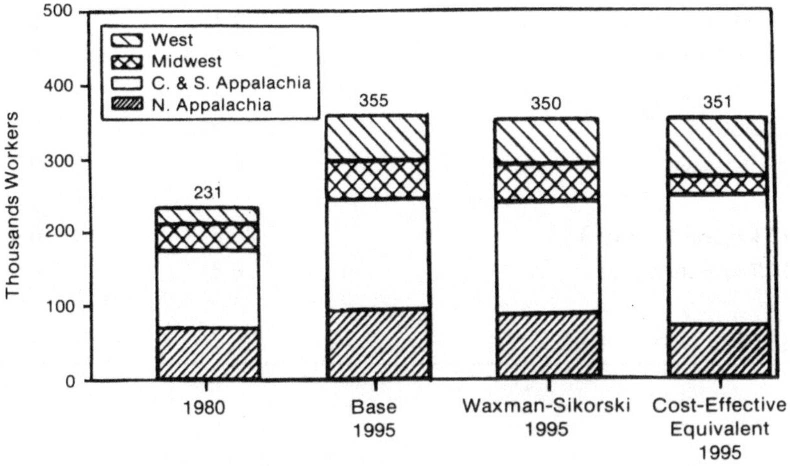

Figure 9b. Shifts in Mining Employment

COST PER JOB SAVED

The cost per job saved under Waxman-Sikorski is substantial.

Table 2

Waxman-Sikorski is a High-Cost Approach for Saving Jobs

	Waxman-Sikorski	Cost-Effective Equivalent	Difference
Change in 1995 Annualized Costs (10^9 Dollars)	4.2	3.2	-1.0
Gross Job Losses (10^3)			
• Job slots	-0.2	-19.5	19.3
• 1980 Miners	0	-12.8	12.8
Annualized Cost Per Job Saved			
• Job Slots			$52,000
• 1980 Miners			$78,000

NEAR-TERM OPPORTUNITIES

There are opportunities to reduce emissions within 5 or 6 years, but these opportunities are substantially less than those available in 10 or more years.

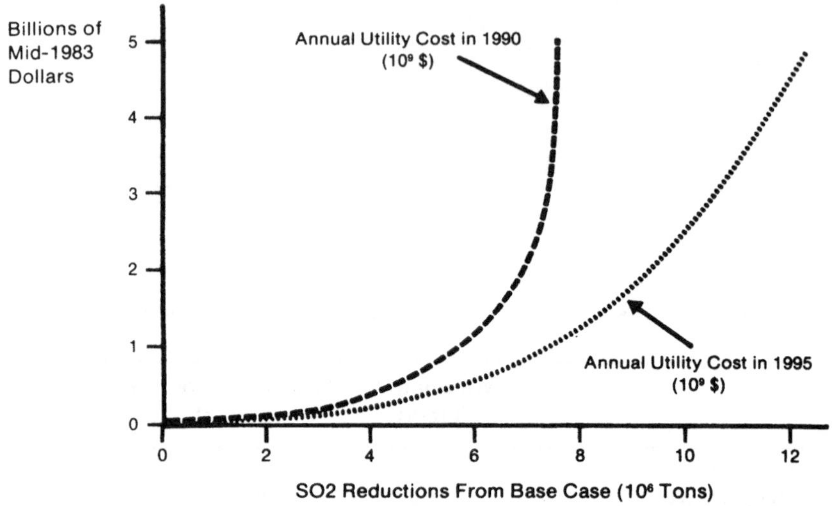

Figure 10. 1990: Cost-Effective Reductions Achievable

Efforts to reduce emissions too much too soon would result in very high costs and much disruption, because capacity constraints on low sulfur coal and scrubber construction would be encountered.

Waxman-Sikorski requirements probably can't be met. The required scrubbers could not be built at assumed costs, if at all, in the time available.

RESPONSE TO HOFF STAUFFER

Larry B. Parker

Analyst in Energy Policy
Congressional Research Service
Library of Congress
Washington, DC 20540

INTRODUCTION

In conducting and reviewing acid rain analyses over the past three years, I have found four basic categories of controversy regarding economic analysis of acid rain: (1) controversy regarding the economic costs versus the anticipated benefits of proposed legislation, (2) controversy regarding application of economic theory as embodied in the various models analyzing acid rain proposals and the perceived reality of utility decisionmaking, (3) controversy regarding implementation of a reduction program, and (4) controversy regarding the long-term integration of proposed acid rain legislation with the rest of the Clean Air Act. ICF Incorporated has been in the forefront of attempts to apply economic theory to the various proposed reductions in sulfur dioxide (SO_2) emissions. Their analyses are widely used by a variety of interest groups on both sides of the policy debate. Their analyses have also been a focus of controversy on the usefulness of economic theory in assessing the economic impact of acid rain controls. Hence, for this presentation, I will concentrate on this category of controversy, and on what I believe to be the healthy discussions which have resulted.

ECONOMIC THEORY AND REALITY: CONTROVERSY IN ACID RAIN ANALYSIS*

As suggested by Hoff Stauffer's presentation, the uncertainty regarding the *direct* cost of acid rain legislation is probably less than the uncertainty with respect to the benefits of such legislation. This is partially due to a lack of definition as to whether such legislation should be considered acid rain legislation, SO_2 reduction legislation, or both. However, as discussed below, cost analyses are now also beginning to experience the same kinds of definitional problems as groups argue about inclusion or exclusion of various *indirect* economic consequences of proposed legislation.

*The opinions expressed in this presentation are solely those of the author and do not necessarily represent those of the Congressional Research Service.

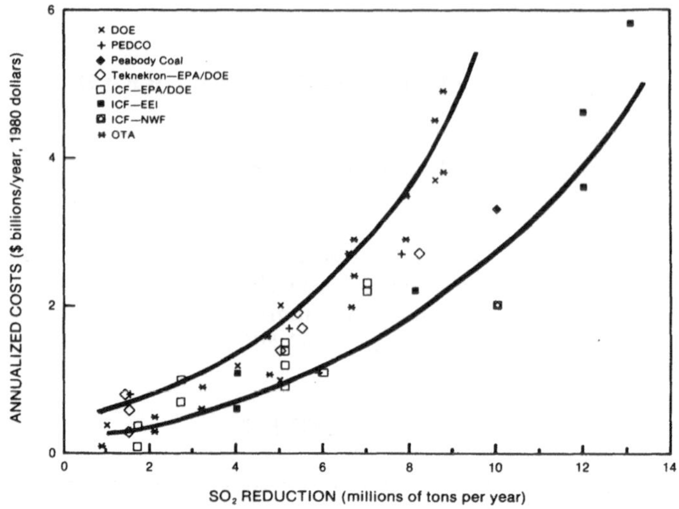

Figure 1. Comparison of Utility SO_2 Control Cost Estimates

Source: Congressional Office of Technology Assessment. 1982. The Regional Implications of Transported Air Pollutants. U.S. Government Printing Office. Washington, DC.

Figure 1 is a plot of analyses done by various models of the coal market and utility industry. As indicated, the models all agree on the upper sloping nature of SO_2 reduction costs. However, the plot also indicates that, besides increasing costs, there is increasing uncertainty among the models as reductions increase, particularly beyond eight million tons. Hence, while an estimate of $3-$6 billion annually for a ten-million ton reduction is a good ballpark figure of the costs, it still represents some uncertainty.

Yet, the controversy over costs is not really between the models, but rather between the models with their $3-$6 billion range, and various utility estimates that a ten-million ton reduction would cost $10 billion plus annually. As discussed by Stauffer, assumptions and presentation explain much of the difference between various cost estimates. As shown in Table 1, an analysis conducted by the Congressional Research Service (CRS) for the Senate Environment and Public Works Committee indicates that in some important cases, utilities, the group which would have to make most of the reductions under proposed legislation, are not accepting the "least-cost" methodology as espoused by the models. This apparent refusal by utilities to accept the strategies suggested by the models to be the most cost-effective raises the first issue in the economic theory versus reality category:

Are "Freedom of Choice" Programs the same as "Least-Cost" or "Cost-Effective" Programs as analyzed by the models?

A "Least-Cost" or "Cost-Effective" program, as modeled by ICF and others, looks at individual powerplants and analyzes a variety of alternative reduction techniques to achieve the most cost-effective reduction. This cost-effective criterion extends to the allocation of the powerplant's reduction requirement, or even the

Table 1

Relative Comparison of Various Cost Studies*

Factor Influencing Cost a/	Study													
	DOE	EEI	EPA	NWF	OTA	TVA	AEP	IPL	NEES	NYPP	PSI	OHIO	SCS	UE
1. Amount of SO_2/NO_x to be reduced	2	3	3	2	1	2	3	2	2	2	1	2	2	2
2. Allocation of removal sources	b/	1	1	1	1-2	2	3	2	2	2	3	3	2	3
3. Scope of allocation	3	3	2	3	3	3	3	3	3	3	3	3	3	3
4. Alternative chosen	b/	1	1	1	1-2	3	3	2	3	3	3	3	3	1
5. FGD costs	2	2	2	2	1-3	1	3	2	n/a	3	2	3	1-3	2
6. Fuel switching costs	2	2	2	2	2	n/a	1	3	c/	c/	2	n/a	n/a	1
7. Accounting method	1	1	1	1	2	1	3	3	n/a	n/a	3	3	1	3
8. Replacement costs	1	1	1	1	1	2	3	3	1	2	2	1	1	1

*NOTE: NO ATTEMPT HAS BEEN MADE TO WEIGH THE DIFFERENT FACTORS. HENCE, COLUMN TOTALS ARE NOT MEANINGFUL.

Source: Parker, Larry B. 1982. Summary and Analysis of Technical Hearings on Costs of Acid Rain Bills. Congressional Research Service, Washington, DC.

a/ Scale for Questions

1. (1) under 10 million tons, (2) 10 million tons, (3) over 10 million tons
2. (1) least cost, (2) emissions ceiling, (3) percentage reduction
3. (1) includes NO_x and industrial emissions, (2) includes utility NO_x, (3) only utility SO_2
4. (1) least cost, (2) significant fuel switching, (3) mostly FGD

5&6.(1) under the CBO estimate, (2) approximately the CBO estimate, (3) significantly above the CBO estimate

7. (1) levelized costs, (2) annualized, but not levelized costs, (3) first-year rate impact
8. (1) no replacement capacity, (2) purchase power, (3) build new facility

state's reduction requirement in interstate trading scenarios. Within certain important constraints, the coal market is assumed to be relatively unhindered and utilities and state governments are assumed to make economically-based decisions.

The analyses done by many utilities suggest that a "Freedom of Choice" bill might not result in a "Least-Cost" or "Cost-Effective" program as outlined by the models. Besides dismissing least-cost allocations and interstate trading as impractical, some of the utility studies indicate that least-cost is not the methodology which they are using to determine reduction strategies in their position papers. Instead, some midwestern utilities are saying that, if feasible, their powerplants would install flue gas desulfurization (FGD) to meet reduction requirements. If FGD is not feasible, the utilities are then saying that the plant is assumed to be retired or fuel switching is considered. As noted by Stauffer, fuel-switching is generally the most cost-effective strategy for reducing SO_2. However, as we see, some utility analyses treat this as the alternative of last resort--a treatment that could become a self-fulfilling prophecy by encouraging the Congress or states to mandate FGD.

Why is this the case? Hoff Stauffer has already outlined the economic advantages of fuel switching over FGD. Table 2 identifies some of the non-market aspects of FGD. As indicated, the primary advantage of FGD, and that most cited by midwestern utilities for employing it, is reduction in high-sulfur coal market dislocations. Against this advantage is a host of disadvantages. These disadvantages have resulted in utilities overwhelmingly choosing fuel switching to meet requirements of State Implementation Plans (SIPs) during the 1970s, as illustrated in Table 3 for Illinois.

Table 2

Concerns Regarding FGD

For FGD	Against FGD
1. State concerns over employment effects of coal switching.	1. Coal switching provides greater reliability and easier implementation than FGD.
	2. FGD requires significant financing.
	3. FGD requires various PUC rate base approvals while coal switching may be passed through on a fuel adjustment clause.
	4. Difficulties of retrofitting FGD to a plant not designed for it.
	5. Waste disposal concerns.
	6. History of FGD development and start-up.

Table 3

Fuel Changes in Coal-Fired Power General Capacity in Illinois
(megawatts)

Total Coal-Fired Power Generation Capacity in 1972		16,424
Converted to Low-Sulfur Western Coal by 1977	6,852	
Converted to Low Sulfur Eastern Coal by 1977	351	
Scheduled to Convert to Low Sulfur Coal	1,673	
Maintaining High Sulfur Coal Combustion	5,798	
Conversions to Oil		430
Retirements by 1977		1,320
Additions since 1972		3,800
Low Sulfur Western Coal	893	
Low Sulfur Eastern Coal	450	
Scrubbers	1,187	
Variances	1,270	
Total Coal-Fired Power Generation Capacity in 1977		18,474

Source: Form 67, Federal Energy Regulatory Commission, 1977. Table from Boyce et al., 1981. Implications of Expanding Coal Production for Illinois' Transportation Systems. Illinois Institute of Natural Resources, Chicago.

In view of the disadvantages presented in Table 2, and the past history illustrated in Table 3, it is clear that the utility analyses are intended to suggest that others will force FGD on the industry. Such a scenario, if correct, would limit the usefulness of least-cost analysis. However, if the past is prologue, then the models of ICF and others are realistic. Given Freedom-of-Choice, history suggests that utilities will use fuel-switching (the least-cost option according to the models) extensively to achieve reductions, resulting in high-sulfur coal market dislocations.

This expressed concern about the high-sulfur coal industry brings up a second issue in the economic theory versus reality controversy:

Is Buying Displaced High-Sulfur Coal Miners Condos in Florida Cheaper than Buying Control Technology to Preserve High-Sulfur Coal Markets?

ICF's analysis of the *direct* cost of a mandated technology versus a cost-effective program indicates that saving *miners'* jobs by mandating technology is very expensive. However, such analyses bring out the definitional dispute over costs I mentioned earlier--that is, indirect costs of coal market dislocations. In the area of job creation and dislocations, economists tend to become amateur sociologists with usually glib discussions or dismissals of the importance of indirect costs. Indeed, at a conference I attended last week, I listened while a respected economist, after expounding on the virtues of buying off the dislocated miners rather than mandating technology, stated that because the indirect costs of dislocations were somewhat uncertain, he just ignored them in his work. May I suggest that this response is not warmly regarded by legislators representing high-sulfur coal miners.

An example of the economic importance of indirect costs of miners' job dislocations may illustrate why such jobs have received an elevated position in the acid rain debate. Figure 2 identifies five counties in western Kentucky which produce about 90 percent of western Kentucky's coal. According to a least-cost analysis of an eight-million ton reduction, these five counties would lose about 40 percent of their production.

What kind of economic impact would this have on the five counties? Tables 4 and 5 provide a 1980 employment and personal income profile of the counties. As indicated, one out of four jobs in the area is in the coal industry. Indeed, it is unclear whether the communities within these counties would exist without coal. For example, the second largest category, services, and the third largest, government (mostly school teachers), are very closely related to supporting the coal industry.

However, perhaps more important from an economic standpoint is the distribution of personal income. As indicated, those mining jobs account for over 40 percent of the personal income within the five counties. The top three categories account for two-thirds of the area's personal income. Hence, in discussing mining jobs, one is discussing more than just individual mining jobs, and comparing them to the cost of FGD. Rather, one is talking about the economic viability of some communities. In this context, discussions of relocation and retraining make little political--or economic--sense.

Hopefully, as the debate continues, economists will stop talking about jobs, and instead, discuss impacts on regional income and wealth which is the real economic issue and something economists are more trained to do.

The flip side of the employment debate is the impact of acid rain on electricity consumers. This third issue in the economic theory versus reality controversy is:

Are Programs Which Are "Least-Cost or Cost-Effective for Society" the same as Programs Which Are "Least-Cost or Cost-Effective for Electricity Ratepayers?"

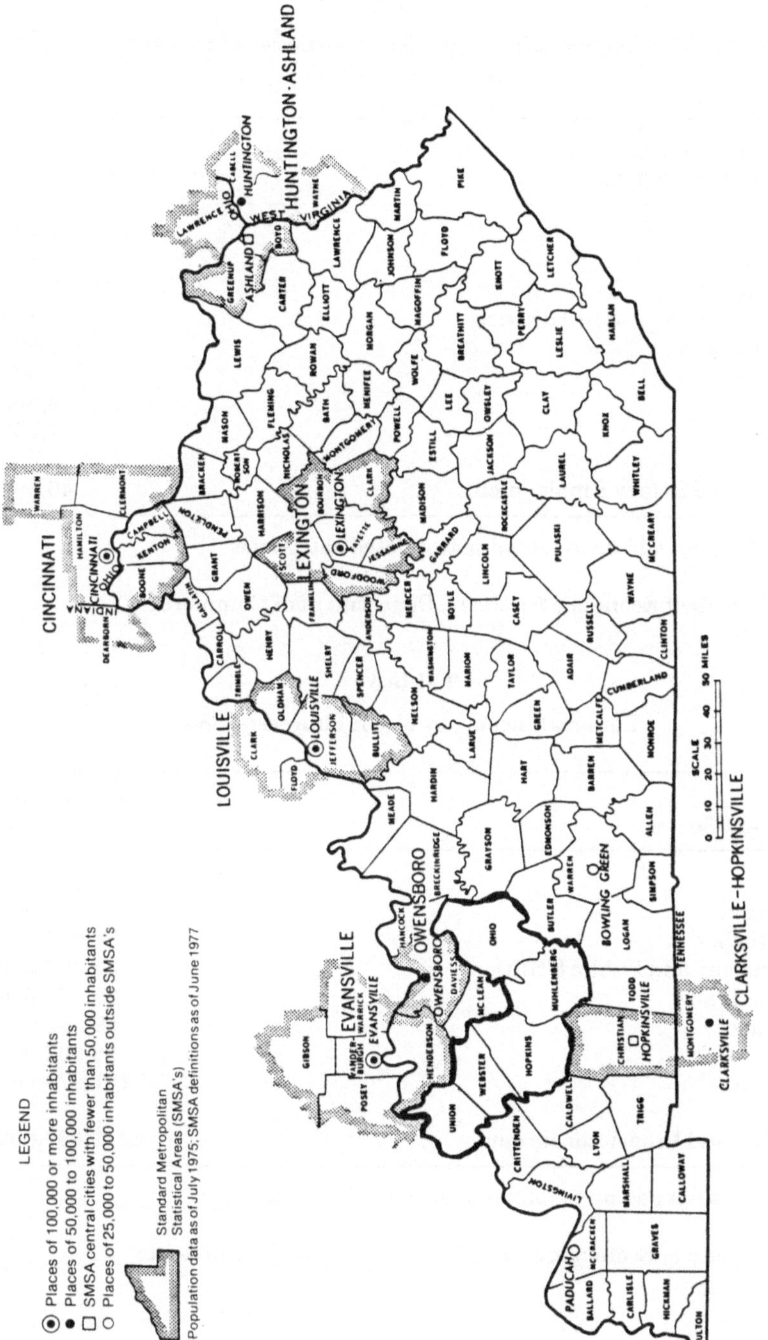

Figure 2. Western Kentucky Counties Examined

Table 4

1980 Employment in Five Counties in Selected Categories
(full- and part-time)

Wage and Salary Employment Category	1980 Employment
Mining	11,121
Construction	1,812
Manufacturing	5,647
Transportation and Public Utilities	2,749
Wholesale Trade	1,508
Retail Trade	5,601
Service	8,082
Government	7,769
Total Wage and Salary Employment	46,663*

*Column does not add up to total due to deleted categories.

Source: Bureau of Economic Analysis, Department of Commerce.

Table 5

Personal Income by Major Sources: 1980

Employment Sector	Labor and Proprietor Income
Mining	$377,375,000
Construction	39,067,000
Manufacturing	84,944,000
Transportation and Public Utilities	62,840,000
Wholesale Trade	29,477,000
Retail Trade	59,112,000
Services	86,362,000
Government	101,979,000
Total Labor and Proprietor Income	$871,699,000*

*Column does not add up to total due to deleted categories.

Source: Bureau of Economic Analysis, Department of Commerce.

I believe that Stauffer's paper has gone far in advancing the discussion of acid rain economics by avoiding discussion of potential electricity rate increases resulting from such legislation. In discussing acid rain, the central economic issue is whether the marginal costs of reducing an amount of SO_2 would be equal to the marginal benefits received, not whether a utility's rates increase 5, 10, or 15 percent over its 1980 rates.

However, the current debate is framed in terms of electricity rates, and indeed, least-cost analyses, such as ICF's work, have been used by some to assert a range of impacts on electricity consumers. This assertion is stretching the least-cost methodology beyond the discussion of cost to society as a whole to the impact on one particular segment of the economy. Such an extension of modeling results has raised serious issues regarding whether electricity consumers would be the recipients of any societal "savings" from a least-cost program, or whether other participants would capture these economic rents.

There are three components of the coal supply system: (1) mining and preparation, (2) transportation, and (3) conversion. Also, at the first and third part of the process, state governments may intervene into the process in the form of severance taxes and public utility regulation. Considering this process in general terms, state severance taxes and increased transportation tariffs probably represent the greatest opportunities to capture rents under acid rain legislation, although some have argued that low-sulfur coal producers may also have opportunities to significantly raise prices.

The competitiveness of coal transportation, particularly rail carriers, has been a source of considerable discussion and debate in recent years. With many of the "Freedom of Choice" bills in the Congress requiring utilities which choose to fuel-switch to do so within five years, questions have been raised as to whether short-term stresses on the low-sulfur coal market and transportation links, along with potential panic buying, would present opportunities for some to constrain supply and capture rents. With the demise of the proposed ETSI pipeline, this issue has become particularly acute with regard to western coal. Unfortunately, inadequate analysis on this issue has been done.

Of course, coal-producing state governments have considerable flexibility in setting severance tax rates. Some states, such as Montana, are currently cashing in on the increased value of its coal from Clean Air Act regulations. In its legislative findings on its current 30-percent severance tax, the Montana Legislature declared that, "Coal in Montana, when subbituminous and recoverable by strip mining, is in sufficient demand that at least one-third of the price it commands at the mine may go to the economic rents of royalties and production taxes." Whether acid rain legislation will encourage other states to adopt a similar stance is unknown.

The timing of reductions and utilities' attitude toward fuel switching will have a major impact on whether individual coal-producing states, rail carriers, or even low-sulfur coal producers would have the opportunity to capture rents. The "reality" of such market imperfections are in need of additional analysis.

CONCLUSION

Assumptions and methodologies used in economic analyses of policy issues have political implications. The history of air quality legislation is one in which political considerations have frequently dominated economic ones. Policymakers have been less concerned with finding the economically "optimal" solution than with addressing issues of regional equity and "acceptable" costs. To the extent systematic sensitivity analysis is conducted on important economic parameters, modeling results assists the Congress in designing acceptable programs and avoiding excessive costs.

However, different analyses done by various interest groups, specifying economic parameters, definitions, and methodologies which favor their political position, results in the confusion we see in the acid rain debate. This result is probably unavoidable, but its impact may be to relegate economics to a minor role in the final policymaking process.

RESPONSE TO C. HOFF STAUFFER

Robert W. Crandall

Senior Fellow
The Brookings Institution
1775 Massachusetts Avenue, N.W.
Washington, DC 20036

As I understood our charge, it is to look at the uncertainties in the costs of potential acid rain control strategies and in all of the assumptions that go into such an analysis. In fact, that turns out to be such a complicated problem that none of the papers dared to delve very deeply into it. I suppose if I had the black box to sell, as Hoff Stauffer does, I would not let you look too far inside it, either. On the other hand, if we really got into some of the details, it would probably put you all to sleep very quickly (except those of you who are here to compete with him in the market).

LOOKING BACKWARD

When one looks into environmental policy, one notices the absence of ex post assessment of existing or past policies. Most of the attention is given to ex ante analysis of proposed programs. You would think from listening to this discussion that perhaps we are embarked on a great new program of controlling SO_2. It is my understanding, and check me if I am wrong, that we have been controlling SO_2 for decades. Yet, we have very little in the way of analyses to show how well we have done, what it has cost, how we might have done it better. To be sure, there are a few cost-benefit analyses in NAS reports written at the behest of Congress, but there is relatively little ex-post analysis of any environmental program. Now, the absence of retrospection does not seem to exist in other areas of policy, even defense policy. One of the reasons for the absence of such evaluations of environmental policy is that we do not have information upon which to base such analyses. We do not have satisfactory data, for instance, on how much we have spent in the past, how it has reduced emissions or effluents, nor how much the air or water has improved in quality. Often, the data on emissions and air quality don't even move in the same direction.

It appears that the information is finally getting better. The EPA is apparently improving the air quality monitoring network around the country. A lot of people, particulary newspaper reporters, do not understand that estimates of national air quality or emissions are often very much "seat of the pants" estimates. The numbers on emissions are hypothetical numbers unlike, for instance, census figures on manufacturing or retail trade.

Let me give an example from EPA's annual emissions estimates. Each year that national emissions estimates are calculated, the estimates are changed for virtually every preceding year. For instance, in December 1978, EPA estimated that 1977 SO_2 emissions were 30.2 million tons, but in September 1982, 45 months later, this had been reduced to an estimate of 28.7 million tons. I don't want to sound too cynical, but it seems to me that we can get an 8-million ton reduction for 1977 by just waiting for another 20 years for the official EPA estimate. And that ought to be the most cost-effective strategy of all. With numbers that are shifting around so dramatically, it is very difficult to do any kind of careful ex-post assessment. And as a result, there is no check on these programs. There is no looking back to determine how well we did in the past.

LOOKING FORWARD

When we get to actually looking forward, as Hoff Stauffer suggested, we are on much more certain grounds in measuring costs than in analyzing the benefits of SO_2 reductions. This does not mean that we cannot place a value on the benefits, but that we cannot be sure that there are benefits. On almost every environmental issue that I have come up against, attempting to calculate an economic value of the benefits is far less difficult than trying to calculate the physical benefits themselves. To what extent has some aspect of human satisfaction been improved--the reduction of disease, premature death, damaged plant life or whatever?

PROBLEMS IN MEASURING COSTS

The cost analyses are subject to less uncertainty, yet even here the uncertainties can be rather large. First of all, one should not think of the costs of control as simply compliance costs on the part of the targeted utility. To an economist and to consumers, costs are the reduction in output in the economy caused by the policy in question. There are many more things that go into costs other than simply building the scrubbers.

One of the things that troubles me a great deal is that ICF has not calculated the indirect costs of SO_2 control. An 8-, 10-, or 12-million ton reduction in SO_2 emissions will drive up electricity costs rather substantially. This, in turn, will lead to a good deal of substitution. We know that a substantial amount of substitution took place in response to the energy crisis in the 1970's. This substitution is not only among the types of fuels within a given utility but also among sources of energy for users of that energy.

People may choose to make what otherwise would be sub-optimal locational decisions because of the increase in the price of electric pwoer. There will be switching among products, which we know has clearly happened as a result of energy price increases in the 1970's. The automobile today is quite different from what it looked like at the onset of the 1973-74 Arab oil embargo. One has to look throughout the system, not simply at the total expenditure by public utilities on scrubbers or the higher prices of coal.

Secondly, simply to calculate engineering costs and then to assume that the scrubbers obtain the reduction that the engineers predict is naive. In the 1960's, the McNamara Commission report on the SST provided an analysis of how many of these aircraft would have to be produced in order to justify the government's investment in the prototype, what they would have to sell for, what the repair would have to be, what the fuel consumption would be, and the value of travelers' time that would be required in order for the SST to be cost-effective in competing with subsonic aircraft.

It turned out that one had to assume zero-percent subsonic flight time in order to justify the government's investment and only then if 500 of these behemoths were sold. One was not told what the McNamara Commission assumed about airport runway lengths for planes landing at 1800 miles per hour! Even then, the government could get a return of only about one-half of the social cost of capital. If 1,000 SST's were sold, an extremely unlikely event, the return would be satisfactory.

Fortunately, the U.S. abandoned this folly but the French and British waded in with disastrous results. One thing that we didn't know about the proposed U.S. SST, that the engineers didn't tell us, was that if we actually produced this aircraft it would have suffered metal fatigue very early in its life. It was so heavily powered because of its weight, that its wings would have fallen off like the Electra of the 1950's. The assumption that the SST would fly for very long periods of time were wrong.

The same sort of problem applies to many engineering analyses. Regularly, the Rand Corporation provides for the Defense Department an analysis comparing ex-post with ex-ante cost assessments of new units of technology for the Air Force (aircraft principally) and they show that the cost overruns and performance shortfalls are the rule rather than the exception in radical new technologies.

These same overly optimistic engineering analyses are likely to be present in the current example of SO_2 control. One cannot assume that scrubbers will produce a reduction of emissions equal exactly to the engineering design. Without careful monitoring at the site, the scrubbers are likely to perform much less well than their theoretical design standards.

Finally, there are a number of other economic considerations that one might want to evaluate. First, what are the dynamic effects of forcing utilities to choose a specific technology now as opposed to waiting for the development of an optimal technology later? Secondly, what are the welfare effects of using a kilowatt-hour tax? This tax imposes social costs of its own that have not been estimated. And finally, how do the utilities behave in response to these constraints? The last of these questions is very difficult. There is no well-settled economics of how utilities allocate resources subject to a regulatory constraint. I think it is safe to say we have a very elegant theory of what happens if regulators set the allowed return above the necessary rate of return. There is a question of whether that theory is right, and a lot of empirical tests of it suggest that it may not be correct. It is not clear that utilities are, even without political constraints, cost minimizers.

One final comment, because Lary brought it up. I think it is a little unfair to say that economists haven't looked at the economic consequences of worker displacement. The Center for Naval Analyses has undertaken a number of studies of what happens to displaced workers in major industries, and they are finding some interesting things. Surprisingly, they find that most workers are able to find new jobs in the same or other geographical areas within a relatively short period of time. Economists don't suggest that dislocation is not a problem. On the other hand, there is a real problem if we take as a goal of economic policy maintaining the distribution of jobs as it is now up to 1995 or 2005. I just returned from the United Kingdom where that has been the policy for the last 20 years. The industrial wage rate there is about $5.00 to $6.00 an hour as opposed to $12 to $13 here. We might ask workers both in the coal mines and outside if that is the sort of policy they want to pursue.

RESPONSE TO C. HOFF STAUFFER: IS ENVIRONMENTAL REGULATION HAZARDOUS TO YOUR HEALTH?

Frederick C. Dunbar

Vice President
National Economic Research Associates, Inc.
123 Main Street
White Plains, NY 10601

BACKGROUND

Where does the public interest lie in regulating environmental quality, workplace safety and hazardous products? The proponents of regulations argue that accidents and toxic exposures pose unacceptable risks to life and health. The opponents usually argue that the costs of regulations are exorbitant relative to their benefits. At first glance, it would appear that the decision to regulate only involves a trade-off between dollars and lives, a seemingly crass, and possibly unethical, comparison.

This view, however, is overly simplified and not realistic. The costs of complying with health and safety regulations can also be gauged in terms of its effects on people's lives. The costs of regulations are passed through to consumers in the form of higher prices and taxes. These consumer and public resources will not be used for other goods and services which improve the public's health, safety and welfare. The correct way to evaluate a regulation is to compare the public welfare with and without the regulation. Such a comparison is difficult, and emotion laden, because the regulations are imposed on activities and products which pose risks to life and health.

However, the health and safety regulations of the past two decades are not alone in this regard. Almost all public and personal decisions have historically and inevitably involved risks to life, limb and health. Some of these decisions are quite obvious attempts to reduce health and safety hazards. For example, local public officials have long been required to determine the best levels of police, fire and public health protection. At the national level, transportation safety and flood control have been long-standing concerns. It has only been a logical extension of an increasingly affluent society that more recently we have asked our elected officials to reduce hazards from the environment, workplaces and consumer products. Individuals also make such decisions in their daily lives, directly in medical care, safety devices, nutrition and indirectly in the way they value various products and services and the places they live and work.

Unfortunately, the fact that these decisions are made all the time does not mean that they are always easy to make. Public officials have had great difficulty balancing the desire to create a safe environment with the reality that resources are inherently limited. This means that creating safer conditions in one area reduces our ability to improve human welfare in another.

This point can be illustrated by reference to environmental regulations. Lower emissions reduce the probability of premature mortality from exposure to pollution. However, environmental compliance also means fewer resources available to consumers, income that goes to higher prices, taxes or more costly substitute goods and services rather than other goods and services which improve human welfare. Some of the alternative uses of these resources include personal expenditures on medical care, nutrition and housing. Others, to be made by public officials with lower tax revenues, include investments in public needs, such as police, fire and health. In addition, lower income and higher unemployment have psychological impacts which represent a variety of health hazards. Clearly, from the standpoint of reduced risks to the public, environmental regulations are not an unmixed blessing.

In the pages which follow, I attempt to show how decisionmakers can compare what the public is giving up if they impose costly health and safety regulations. This approach is aimed at determining whether the public is better served by less strict (i.e., less costly) regulation. Briefly stated, the approach estimates the expected lives lost because of lower effective income caused by regulatory compliance relative to the expected lives lost from exposure to health and safety hazards.

EFFECT OF REDUCED INCOME ON HEALTH

It has long been recognized that increasing the standard of living improves human health. This goes well beyond the use of national wealth for public health projects. Personal well-being, both physical and mental, is affected by the level and stability of income. When income or its security falls, individuals cannot afford to maintain their health and safety at existing levels and, perhaps even more important, mental stresses mount which lead to psychopathic behavior and physical deterioration. The result is a higher mortality rate.[1]

One way to view the costs of health and safety regulations is that it reduces the aggregate, effective income of consumers by the costs of compliance because this is the increase in the cost of living which will be borne by consumers. For example, pollution abatement and control expenditures amounted to 2.5 percent of personal income in 1981 (the latest year for which figures are available) and of these, air pollution control was responsible for 1.2 percent of personal income.[2] A large portion of these expenditures are promoted on the basis of their health effects. These values

[1]One of the first studies of this effect is by Kitagawa and Hauser (1973). Another major study and research review is by Brenner (1976) which he then updated and published (Brenner, 1984).

[2]Pollution control and abatement expenditures were $60.326 billion in 1981 of which $29.5 billion was for air quality control (Statistical Abstract, 1984). Personal income in 1981 was $2,435.0 billion (Economic Report of the President, 1984).

exclude expenditures for increasing product and workplace safety, as well as those caused by banning entirely some products. These measures are based on the same goal of improving public health and safety regulations; the costs of these regulations are not known.

Though either a 1.2 or 2.5 percent decline in income will not have catastrophic health effects, it will be adverse. We can make an estimate of its effects by using recent results from the literature on epidemiology. There are now a large number of studies which correlate various environmental and socio-economic indicators with the risks of mortality and sickness; much of this research was motivated by the need to evaluate air pollution policy.[3] It is common in such studies to include many factors affecting health so as to isolate the effect of, say, environmental policy. The effects of income, unemployment and poverty levels are often included in the relationships which predict mortality or health risks.

An important example is the work of Dr. Lester Lave whose epidemiology research is perhaps the most often cited in the literature. Lave's early work culminated in a report published in 1977 (Lave and Seskin, 1977). In response to a number of reviews of his work, Lave (more correctly, Chappie and Lave) recently updated his analysis incorporating most of the suggestions of his critics (Chappie and Lave, 1982). In so doing, Chappie-Lave uses statistical techniques (called multiple regression analysis) to estimate the effect of many different factors on mortality rates across more than 100 U.S. metropolitan areas in the 1970s. They specify about 40 different relationships which can potentially explain mortality rates. All of these relationships include environmental factors (mainly air pollution) and several or more socio-economic factors which affect health. About 34 of the relationships include median family income and the remainder include other indicators of income such as the percent of the population below the poverty line.

Whenever Chappie-Lave isolated the effects of family income on mortality, it was highly significant.[4] Table 1 summarizes the effects of income across a number of different relationships estimated by Chappie-Lave. In Table 1, I present two ways of representing their results. The first (Column 4) shows the effects on the mortality rate (number of deaths per 100,000 of population) of increasing median family income in an urban area by one dollar. Another way is to compute an elasticity; this elasticity is defined as the percent change in mortality rates divided by the percent change in median family income. An elasticity is a useful way of summarizing information because it abstracts from problems of inflation, and perhaps other socio-economic changes, which have occurred since the time of the data in Chappie-Lave's sample

[3]For a brief review of existing air pollution studies see Freeman (1979).

[4]More precisely, income significantly affected mortality rates except on a few occasions when percent of the population which is college graduates was included as an explanatory variable. The reduction in significance of income on these occasions is apparently due to a statistical problem called multicolinearity; when this is present, it is not possible for the data itself to isolate the true effect of income on the mortality. The authors conclude, as I do, that income rather than college education is the appropriate causal factor on mortality rates.

Table 1

Summary of Estimated Effects of Income on Mortality Rates from Chappie and Lave

Equation Number	(1) Dependent Variable[1]	(2) Sample Means: Deaths per 100,000 Population (1974)	(3) Sample Means: Median Family Income (1972 Dollars)	(4) Estimated Coefficient: Change in Deaths per 100,000 Caused by a $1.00 Increase in Median Family Income	(5) Elasticity: Percent Change in Mortality Caused by a One Percent Change in Income (3)×(4)÷(2)	(6) Regression Specification
2-4	Total Mortality Rate	867.4	9,859.0	-.020	-.277	Includes effects of pollution, age distribution, race and population density
3-5	Natural Mortality Rate	793.5	9,872.3	-.036	-.448	Similar to 2-4 but also includes effects of smoking, alcohol use and industry mix of employment
4-3	Natural Mortality Rate	793.5	9,872.3	-.024	-.299	Similar to 2-4 but also includes effects of smoking, alcohol use and weighs urban areas by size
5-6	Total Mortality Rate	1,092.5	9,235.9	-.032	-.287	Similar to 2-4 but also includes effects of smoking, alcohol use and only includes cities rather than entire metropolitan areas
6-9	Total Mortality Rate	865.9	9,872.3	-.036	-.410	Simultaneous equation model determining physician supply and demand; includes variables from 4-3 and nutrition and medical care
7-3	Total Mortality Rate	865.9	9,872.3	-.108	-1.231	Similar to 6-9 but includes additional nutrition variables

[1]Total Mortality Rate is all causes of death; Natural Mortality Rate is nontraumatic and includes all causes of death except accidents, homicides, suicides and other external causes.

Source: Mike Chappie and Lester Lave, "The Health Effects of Air Pollution: A Reanalysis," Journal of Urban Economics 12 (1982) pp. 346-376, Table 6.

(1974). Inflation would affect the estimated coefficients but would usually have only de minimus effects on the elasticity.[5]

These elasticities are presented in Column 5 and are probably higher than most people would have thought. Specifically, they indicate that a 1 percent increase in median family income reduces the mortality rate anywhere from .28 to 1.23 percent.

The range in the elasticities is relatively high showing that the observed relationship between income and mortality is sensitive to other factors which are included or excluded in the regression analysis. In particular, the income elasticity tends to increase as more variables are included. This means that family income is correlated with other factors which increase mortality rates and by including these factors the effect of family income is made more significant. The most important of these other factors appears to be the proportion of the work force in manufacturing and the nutritional composition of diet. Between these two, I think it is important to isolate the effect of family income from manufacturing jobs, but less important to separate income from nutrition. Typically, higher income areas have a higher proportion of the work force in manufacturing. This latter effect increases mortality independent of any other changes in income because of workplace and other environmental hazards caused by heavy industry. Consequently, the estimated effects of income without controlling for the proportion which comes from manufacturing would underrepresent the true effect of income on reducing mortality.

Before pursuing the implications of the Chappie-Lave results, it is worthwhile comparing them to two other results found in the literature. The first of these is the work of Dr. M. Harvey Brenner of the Johns Hopkins University who has spent over a decade examining the relationships between aggregate economic indicators such as personal income and unemployment rates, and social pathology such as mortality rates, illness and criminal behavior. Like Lave, Brenner uses regression analysis. Unlike Lave, however, Brenner analyzes time-series data spanning the years 1935 through 1980. Brenner's most recent study, the results of which are described below, uses data from 1950 through 1980.

Brenner presents the results of 45 regressions in his most recent monograph (Brenner, 1984 at 66-99 and 95-111). These regressions relate various dependent variables such as mortality from homicides, mental hospital admissions, or mortality by age cohort to a set of independent variables reflecting income, unemployment, consumption of cigarettes and alcohol, and changes in demographics. Of these, the most important for our purposes is a single regression which estimates the effect on the total mortality rate of trends of real per capita disposable income (among other explanatory variables). The elasticity of mortality with respect to income can be derived from results presented by Brenner which are summarized in Table 2. The

[5]The use of elasticities is fairly common in epidemiology. For example, Lave typically uses the elasticity of mortality with respect to pollution levels to compute the health benefits of pollution control (Lave and Seskin, 1977 at pp. 217-221 and Chappie and Lave, 1982, at p. 347). This is the same approach used by Freeman (1979 at pp. vii-viii).

Table 2

Brenner's Estimate of the Change in the Incidence of Trauma Caused by a 10-Percent Change in Disposable Per Capita Income (Based on 1980 Population)

Trauma	(1) Total Incidence in 1980	(2) Increase Caused by a 10-Percent Decline in Income	(3) Percent Increase[1] ((2) ÷ (1)) × 100
Total Mortality	1,986,000	201,850	10.2
Cardiovascular Mortality	1,035,250	150,631	14.6
Cirrhosis Mortality	30,066	1,172	3.9
Suicide	27,640	1,066	3.9
Imprisonment	304,332	7,964	2.6

[1]Corrected from Brenner's original table.

Source: M. Harvey Brenner, "Estimating the Effects of Economic Change on Natinal Health and Social Well-Being" (A Study Prepared for the Use of the Subcommittee on Economic Goals and Intergovernmental Policy of the Joint Economic Committee of Congress) Washington, D.C.: Government Printing Office: 1984, p. 69.

elasticity is unity;[6] this is at the high end of the range found by Chappie-Lave. Brenner's results are quite robust. The statistical significance of income on mortality is extremely high and this level of significance is typically found when the components of mortality, i.e, mortality by cause and age cohort, are regressed on income.

Another study which attempted to quantify the effects of income on mortality was that of Professors Ralph L. Keeney and Detlof von Winterfeldt (1983). The authors use the results of Kitagawa-Huaser's earlier research on mortality rates by social and income class. From 1960 data, Kitagawa-Hauser divided demographic groups into income quintiles and computed mortality rates for each group. Keeney-von Winterfeldt fit a smooth curve through these data points; the curve relates mortality rates to level of family income. The slope of the curve at any point has the same interpretation as a linear regression coefficient. However, because the curve is nonlinear it shows that a unit change in income for low-income families has a higher

[6]Note that Brenner makes an arithmetic error in one part of his text which implies that the elasticity is one-tenth of the true estimate. Table 2 presents corrected results. From the text of Brenner's monograph, it is clear that he has estimated an elasticity of unity (Brenner, 1984 at 3 and 69).

Table 3

Estimated Effects of Household Income on Mortality Using Keeney and Von Winterfeldt's Analysis of the Kitagawa and Hauser Studies

(1) Income Class (Dollars)	(2) Change in Mortality Rate per 100,000 From Loss of \$100 Household Income	(3) Total Number of People (Millions)	(4) Ratio of Mortality Rate for Income Class to Total Mortality Rate	(5) Percent Change in Income (100÷(1))×100	(6) Percent Change in Mortality Rate (2)÷((4)×954.7))×100	(7) Elasticity of Mortality Rate with Respect to Household Income (6)÷(5)
6,000	31	45	1.25	1.67	2.60	1.56
21,000	9	135	0.94	0.48	1.00	2.08
55,000	3	45	0.80	0.18	0.39	2.17
Weighted Average:[1]						
\$24,800	12.2	--	0.97	0.66	1.20	1.99

[1]Weighted by total number of people in column (3).

Sources:

Columns (1),(2),(3),(4): Ralph L. Keeney and Detlof von Winterfeldt, "A Methodology to Examine Health Effects Induced by the Compliance Activities and Economic Costs of Environmental Regulation of Power Plants" (Palo Alto: Electric Power Research Insitute, December 1983) pp. 36 and Figure 4. DRAFT FOR LIMITED CIRCULATION.

Column (6): Total Mortality Rate for 1960 (954.7 per 100,000) U.S. Bureau of the Census, *Statistical Abstract of the United States: 1984* (104th Edition) Washington, D.C., 1983, p. 78.

Table 4

	(1) Elasticity of Mortality Rate with Respect to Personal Income	(2) Expected Number of Deaths Caused by a Reduction of Income of 1.21 Percent in 1981 ($29.5 billion reduction) $(1) \times 1.21 \times 1{,}987 \div 100$ (Thousands)
Chappie-Lave	.45	11
Brenner	1.02	25
Keeney-von Winterfeldt (Kitagawa-Hauser)	1.99	48

effect on mortality than does a unit change in income for either middle or high-income families--a not unexpected result.

Table 3 summarizes the Keeney-von Winterfeldt analysis of the Kitagawa-Hauser cross-tabulations. From their summary, it is possible to compute an elasticity of mortality with respect to family income. This computation is performed in Table 3 resulting in elasticity equal to two, a value higher than found in the other studies reviewed above.

As an example of how these findings can be applied, let us consider the potential expected loss of life from a 1.2 percent reduction in personal income. Recall that this is the proportion of personal income which was spent on air pollution control in 1981. Using an estimate of the elasticity of mortality which is in the middle of the range found by Chappie-Lave, and using the results from other studies, the loss of life from this income reduction can be estimated at between 11,000 and 48,000 lives.[7] Table 4 demonstrates this in detail.

Interestingly, Freeman's highly publicized study concluded that the best estimate of the reduction in mortality from air pollution control was on the order of one percent for the U.S. metropolitan population or, for 1981, about 14,000 lives saved.[8] There are compelling reasons to believe that Freeman overestimated the

[7]This estimate uses the total deaths from all causes for 1981 which was 1,987 thousand (Statistical Abstract-1984, p. 75).

[8]Freeman (1979 at pp. v and viii) uses an elasticity of mortality with respect to air pollution of 0.05 and presumes that air quality had improved 20 percent. This yields a one-percent improvement in the mortality rate from air quality regulation.

number of lives saved. (See, for example, National Economic Research Associates, Inc., 1980 at 6-12 and 6-24.) Even assuming arguendo that this estimate is correct, the number of lives saved could well be less than the expected number of lives lost as a result of the reduction in personal income caused by the regulations.

This estimate alone is not a sufficient argument against these regulations; a cost-benefit analysis should be performed on individual regulations where benefits take account of effects in addition to health impacts, but the finding is nonetheless provocative. As health and safety regulations deal with hazards that become especially costly per unit risk reduction, because many of the most obvious and least costly hazards are controlled first, the regulations could become counterproductive to the nation's public health.

VALUE OF A STATISTICAL LIFE

In performing cost-benefit analyses of regulations, perhaps the most controversial subject is the value which is to be placed on the expected lives to be saved from the regulatory action. Note that cost-benefit analysis does not attempt to place a value on any particular human life. To do so is beyond the realm of economic analysis. Rather, what past studies have attempted is an assessment of the amount which individuals would be willing to pay to avoid risks to their lives. As mentioned before, people make choices in their everyday lives between income and health or safety. It is then possible to infer the average value which individuals place upon specific health and safety risks.

Most such studies rely on information from the labor market. It has long been known that people require higher wages to accept higher employment risks. Carefully designed statistical analyses can determine the increase in income associated with riskier occupations. This then becomes an estimate of the willingness to pay to accept lower risk. For example, Freeman's review of this literature produced an estimate of $1,000 per year as the payment required to avoid a one in 1,000 chance of mortality. This led him to use $1 million (in 1978 dollars) as the value of statistical life (Freeman, 1979 at IX).

More recently, the U.S. EPA issued final guidelines for doing cost-benefit studies which value a statistical life in the broad range from $400,000 to $7 million (*Inside EPA*, January 6, 1984). These estimates were derived from studies of the relationship between risk in the workplace and wages.

The above-cited results on mortality caused by lower income can be used to check commonly used value of life estimates. The following table shows the value of a statistical life computed by dividing into $29.5 billion the expected number of lives lost by reducing national income by this amount. This gives an estimate for 1981. These values are inflated to 1984 using the increase in the Consumer Price Index from 1981 to 1983 and assuming that inflation will be five percent in 1984 (Economic Report of the President, 1984 at 279).

Table 5

	Value of a Statistical Life Implied By the Elasticity of Mortality With Respect to Personal Income	
	(1) 1981 (millions of 1981 dollars)	(2) 1984 (millions of 1984 dollars)
Chappie-Lave	2.7	3.1
Brenner	1.2	1.4
Keeney-von Winterfeldt (Kitagawa-Hauser)	0.6	0.7

Though Freeman's use of $1 million for each statistical life saved appears to be supportable from this approach, the high end of EPA's range is not. Specifically, by implementing regulations which have a favorable cost-benefit balance only when a statistical life is valued at, say, $7 million, the EPA could be costing two to 10 more consumer's and taxpayer's lives than they are intending to save.

CONCLUSION

Over the past two decades, Congress has held the view that health and safety risks in the environment and work place should be reduced no matter what the cost. These attempts to create a zero-risk society are feckless enterprises. The resources mandated to control risks in one area cannot be used elsewhere--the other uses of these resources include the provision of goods and services which improve human health and well-being in other areas. Until Congress and the public grasp the dimensions of this tradeoff, we run the risk of environmental, health and safety regulations which unintentionally do more harm than good.

REFERENCES

Brenner, M.H., 1976, "Estimating the Social Costs of National Economic Policy: Implications for Mental and Physical Health and Criminal Aggression" (Report to the Congressional Research Service of the Library of Congress and the Joint Economic Committee of Congress), Washington, DC: Government Printing Office.

Brenner, M.H., 1984, "Estimating the Effects of Economic Change on National Health and Social Well-Being" (A study Prepared for the Use of the Subcommittee on Economic Goals and Intergovernmental Policy of the Joint Economic Committee of Congress), Washington, DC: Government Printing Office.

Chappie, M. and Lave, L., 1974 and 1982, "The Health Effects of Air Pollution: A Reanalysis," *Journal.of.Urban Economics* 12:346-376.

Crocker, T.D., September 23, 1980, before the U.S. Senate Small Business Committee and U.S. Senate Committee on Environment and Public Works.

Energy Information Administration, June 1983, "Impacts of the Proposed Clean Air Act Amendments of 1982 on the Coal and Electric Utility Industries," Washington, DC.

Freeman, A. Myrick, III, December 1979, "The Benefits of Air and Water Pollution Control: A Review and Synthesis of Recent Estimates," a report prepared for the Council on Environmental Quality, Bowdoin College, Brunswick, ME.

Inside EPA: Weekly Report, January 6, 1984, Vol. 5, No. 1, pp. 10-13.

Keeney, R.L. and von Winterfeldt, D., December, 1983, "A Methodology to Examine Health Effects Induced by the Compliance Activities and Economic Costs of Environmental Regulation of Power Plants," Electric Power Research Institute, Palo Alto, CA.

Kitagawa, E.M. and Hauser, P.M., 1973, "Differential Mortality in the United States: A Study in Socioeconomic Epidemiology," Harvard University Press, Cambridge, MA, Chs. 4, 5, 7.

Lave, L.B. and Seskin, E.P., 1977, "Air Pollution and Human Health," Johns Hopkins University Press, Baltimore, MD.

National Academy of Sciences, 1983, "Acid Deposition: Atmospheric Processes in Eastern North America," National Academy Press, Washington, DC, at 7.

National Economic Research Associates, Inc., 1984, "Acid Rain Control Proposals: Are They Worth It?," *Energy: Annual Statistics Issue, 1984*, IX:2:28.

National Economic Research Associates, Inc., November 1980, "Cost-Effectiveness and Cost-Benefit Analysis of Air Quality Regulation," White Plains, NY.

Stauffer, H., Jr., December 5, 1984, "Cost and Coal Market Effects of Alternative Approaches for Reducing Electric Utility Sulfur Dioxide Emissions," *Conference on Acid Rain: Economic Assessment, Acid Rain Information Clearinghouse.*

U.S. Bureau of the Census, "Statistical Abstract of the United States: 1984" (104th edition), Washington, DC: 1983, p. 214.

U.S. President, "Economic Report of the President-1984," Washington, DC: 1984, p. 243.

CONFRONTING THE ASSUMPTIONS AND UNCERTAINTIES: ASSESSING BENEFITS

ESTIMATES OF ACID DEPOSITION CONTROL BENEFITS: A BAYESIAN PERSPECTIVE

Thomas D. Crocker*

Professor of Economics
University of Wyoming
Laramie, WY 82071

INTRODUCTION

Most conceptions of rationality require that intelligent choices be grounded in beliefs about consequences for personal or collective values. As March (1978) argues, actual human behavior often follows other logics having their own claims to intelligence; nevertheless, the choices that persons, institutions, and societies make are frequently guided by calculations of the benefits and the costs that alternative actions are expected to bring. These calculations involve two kinds of conjectures about the future, conditional on current actions. The first conjecture is a statement of belief about probable future world states; the second is an estimate of probable future expressions of preferences. Those who make the calculations combine the conjectures to produce weighted averages of the values of the identified alternatives, where each weight is a degree of belief that a particular set of consequences will be realized. The expected values that result are supposed to inform the decisionmaker's choice, though not to absolve him from accountability.

It seems reasonable that central decisionmakers in North American democracies need not respect individual beliefs about future possibilities in the same manner as they respect individual tastes. Tastes may be refined or vulgar; still, it may be held that a person's tastes must be respected. Indeed, the welfare economics that forms the analytical basis for the above-mentioned conception of rationality presumes that individuals' tastes are to be heeded. Beliefs, however, can be wrong, whether they refer to future world states, to future expressions of tastes, or to their respective probabilities. It is hard to understand why a government should base its decisions on mistaken beliefs. At this juncture, federal acid deposition policymakers seem to listen to those individuals and group spokesmen who believe that states and tastes are such that further control of acid deposition precursors would most likely bring forth economic costs in excess of benefits. The consequences of this decision will be experienced within the next few decades. Additional information will then have become available, and the outcomes and taste expressions resulting from the currently favored policy prescription to maintain the status quo will have been

*Richard M. Adams has provided some useful comments.

clarified. One can then examine the discrepancies, if any, between policymakers' apparent current beliefs and ultimate realizations. In this paper, I shall argue that these beliefs wrongly downgrade the benefits of acid deposition control, and probably cause much research into the effects of acid deposition upon natural and human systems to be misdirected. Nonetheless, in order to remain faithful to the perspective that I adopt, I must admit that it might be my arguments rather than current policymaker beliefs which will finally be judged wrong.

PROVISIONAL PERSPECTIVES ON CONTROL BENEFIT MAGNITUDES

Voluntary exchange is the core concept of economic efficiency. Clearly, no one will engage in exchange unless they expect to be made better off. Pareto (1927) was the first to exploit the concept to solve the problem of weighing one person's welfare against another's. His strikingly simple criterion states that outcome A is preferred to outcome B if, by the affected individuals' own evaluations, no one person's welfare could be enhanced without making someone else worse off. Whenever exchanges are not subject to market allocation processes, benefit-cost analysis provides policymakers with a measure of the incomes the gainers and losers from a policy will consider to be equivalent to their respective gains and losses, given their initial wealths. In the practice of benefit-cost analysis, the Pareto-criterion has been reduced by Hicks (1939) to inquiring whether those who would gain from a policy alternative could, in principle, compensate the losers and still have some residual gain.

A thorough benefit-cost analysis of acid deposition control requires three kinds of information: (1) the differential changes across space and time that acid deposition control causes in production and consumption opportunities; (2) the responses of input and output market prices to these changes; and (3) the adaptations that emitters and receptors can make so as to minimize losses or maximize gains from changes in production and consumption opportunities and in relative prices. Natural science investigations must be the primary source of information for the first facet. Evaluating the outcomes caused by the latter two facets is the economics portion of benefit-cost analysis. If accurate information on economic consequences is desired, some formal analysis of the emitter and the receptor decision processes that underlie the last two issues is inescapable. The complete concept of control benefits embodies both physical and biological changes in the entities of interest and the responses of individuals and institutions to these changes.

In spite of the high levels of concern the public and the media express about acid deposition, only two published estimates of the aggregate negative economic consequences of its current impacts appear to exist. The Commission on Natural Resources (1975, p. 624) chose $500 million annually. It characterized this choice as "arbitrary". The claims made in Crocker's (1980, 1983) upper bound estimate of $5 billion in 1978 for existing economic activities in the eastern third of the United States were somewhat less restrained, but, as ICF Incorporated (1983) has pointed out, the methods and data bases employed can readily be viewed as inadequate. Inadequacy, of course, implies that enough potential biases are present to cause, given one's objectives, some standard of rejection to be met.

Two opportunities arise to apply a standard of rejection. First, one might refuse to accept the hypothesis that acid deposition is a significant cause of observed changes

Table 1

Estimates of 1978 Maximum Economic Losses Caused by Acid Deposition in the Eastern Third of the United States[a]

Effects Category	Maximum Losses (1978 Dollars)
Materials	$2.00 billion
Forest ecosystems	1.75 billion
Direct agricultural	1.00 billion
Aquatic ecosystems	0.25 billion
Others (health, water supply systems, etc.)	0.10 billion

Source: Crocker (1980)

[a]Estimates are for the potential total benefits due to the complete elimination of acid deposition effects.

in the condition of natural ecosystems and human artifacts. The estimates in Crocker (1980, 1983) are predicated on acid deposition being a significant cause.[1] Or, if one prefers to shift the burden of proof, they are predicated on the fact that no one has yet shown that acid deposition is *not* a significant cause. Second, given that acid deposition is such a cause, one might reject the natural science and economic data bases and methods used to assess the economic consequences of the changes. In the rest of this section, the original Crocker (1980, 1983) calculations used to treat this second question are briefly scrutinized and partly updated. The reader may then revise his perspective accordingly on the gravity of the biases present.

The basic procedure employed in Crocker (1980, 1983) was to obtain inventory counts of existing activities that might be affected by acid deposition, to review and synthesize the available natural science literature dealing with the impact of acid deposition on the flow of services from these activities, and then to multiply the flow changes by invariant prices. The stocks of existing activities (e.g., materials, outdoor recreational days, etc.) were assumed to be unchanged by variations in acid deposition levels. Thus, apart from generous leavenings of the author's professional judgment and experience, that which was earlier called the economic analysis portion of any benefit-cost exercise was neglected; instead, the first facet of such an exercise, the natural science and accounting facet, was emphasized.[2] The results by effects category are presented in Table 1.

[1]See Smith (1981) and Crocker (1984) for discussions of the tradeoffs involved in posing and providing "acceptable" answers to this first question.

[2]If time and research resources are scarce, the wisdom of this emphasis is highly debatable. Crocker (1982), drawing upon original work of Adams et al. (1982), reviewed the structure of economic gains from ambient oxidant reductions to the producers and consumers of 14 annual crops grown in 4 southern California subregions. Of the 56 (14 × 4) cases, the crop output levels in 29 of the cases after price effects and producer adaptations had been accounted for exceeded in absolute terms the triggering biological dose-response effects by a factor of 2 or more. At least in this instance, a focus limited to biological yield effects would have provided a highly inaccurate picture of the effects of air pollution upon actual crop outputs.

Materials Effects

Materials impacts were assigned the highest pecuniary value, even though Haagenrud's and Atteraas' (1980) study on zinc and steel corrosion rates appears to remain the only investigation that includes an explicit term for acid deposition. The stated figure of $2 billion annually in Crocker (1980, 1983) was developed for the United States by applying a set of strong assumptions to a series of Swedish studies that indicated 1970 annual corrosion losses from all sources in Swedish households of about $25 per capita. A reading of the materials decay literature produced a judgment that about one-eighth of observed annual U.S. corrosion and paint deterioration was due to SO_2 and its oxidation on moist surfaces. This judgment, the extrapolation assumptions, and the Swedish household per capita figure yielded the $2-billion dollar estimate.[3] Because the estimate did not include the manufacturing sector, nonmetallic building materials such as stone, nor the losses in unique private and public structures and materials generally thought of as "cultural heritage," it was thought to be conservative.

The conservative nature of the estimate is readily seen when some rough calculations are performed for the manufacturing sector. According to the U.S. Department of Commerce (1980, Table 1444), the 1979 values of the stocks of equipment and structures in the U.S. manufacturing sector were respectively $365 billion and $160 billion. Reported replacement life was 12 years, on average. Equipment purchases in 1979 were $46 billion, and investment in new structures amounted to $14 billion. Assuming a 10-percent discount rate and a reduction in acid deposition levels that extends average replacement life by 6 months (4.2 percent), the present value of the reduction in acid deposition induced damages to the manufacturing sector's 1979 capital stock would be $5.7 billion. Assuming that 1979 investment magnitudes in the capital stock would be continued, the benefits caused by reductions in materials damages to this annual investment would be approximately $0.9 billion per year, or, in present value terms, $8.9 billion over 50 years in the manufacturing sector alone.[4]

The opinion that the $2-billion estimate is conservative is reinforced by the recent conclusion of Greenfield et al. (1983, Table 4-3) that an annual average concentration of only 50 μg/m of SO_2 in Boston, Massachusetts, would in 1983 have caused losses in paint erosion alone amounting to between $9.92 and $20.30 per person. These figures were not lifted from studies of Swedish households, but rather were derived from the detailed Stankunas et al. (1983) field inventory of painted

[3]The key extrapolation assumptions were: (1) that per capita mixes of household building materials in Sweden and the Eastern U.S. were similar, and (2) corrosion and paint deterioration rates were also similar.

[4]For example, suppose a machinery/equipment item with replacement cost C-dollars would be replaced in 10 years with pollution level A_1 and in 12 years with a lower pollution level A_0. The present value of economic benefits attributable to a policy which lowers pollution levels from A_1 to A_0 is $C(1+r)^{-10} - C(1+r)^{-12}$, where r is the discount rate. In general, given a replacement life of t-years for an article, a pollution-level change of ΔA which <u>increases</u> replacement life by Δt years cause economic benefits (B) attributable to ΔA to be:

$$B(\Delta A) = C\left[\frac{1-(1+r)^{-\Delta t}}{(1+r)^t}\right].$$

surfaces in the Boston area, and from dose-response functions set forth in Haynie et al. (1976) and McCarthy et al. (1981).

Forest Ecosystem Effects

Of all the suspected effects of acid deposition, those on forest ecosystems seem the most confusing. Competing hypotheses without attached probability statements abound, and observable symptoms consistent with each hypothesis have frequently been noted. General classes of hypotheses for acid deposition induced forest decline include soil effects, direct foliage effects, and increasing predispositions to stress, but within each class there are numerous complex, non-nested and possibly incoherent subhypotheses (Johnson, 1983), each of which can be given specific empirical content in a variety of ways.[5] Acid deposition need not be the only cause of the observed symptoms. Nevertheless, whatever the causes, Jacobson (1983), Tomlinson (1983), and others agree that many commercially and aesthetically important tree species in the U.S. have been "dying back" and declining for several decades.

In Crocker (1983), Jonsson's (1976) position that no good reason exists not to attribute observed reductions in forest growth to acid deposition was adopted. Accordingly, the 5-percent average annual growth reduction in eastern U.S. forests that the Panel on Nitrates (1978, p. 577) attributed to acid deposition was used. A simple multiplication of this growth reduction by a U.S. Forest Service (1978) weighted average of 1977 stumpage prices (in 1978 dollars) resulted in an estimate of nearly $600 million annually in lost timber production. Based on an extrapolation from a Calish, et al. (1978) study of a western Douglas fir forest, more than $1 billion annually was added to this sum for losses in forest outdoor recreation, water storage, wildlife habitats, and other forest services which depend on the condition of the standing vegetation. These latter estimated losses were equivalent to average annual value reductions of about $4 per acre, given the approximately 300 million acres of commercial or productive forest land that the U.S. Forest Service (1978) includes in the states east of the Mississippi and in Minnesota.

Five years after the above calculations on forest impacts were performed, no information has yet appeared which, by normal standards, justifies rejection of the hypothesis that acid deposition is the cause of observed annual growth reductions in U.S. forests. Recent reports such as Johnson (1983) make it seem plausible that the annual growth reduction exceeds 5 percent. Admittedly, however, most of the alternative hypotheses used to explain forest decline, such as drought in the 1960's, have not yet been rejected either. In fact, a good deal of scientific ingenuity has been displayed in finding new non-nested, plausible sources of explanation for the observed growth reductions. Indeed, as soon as some predecessor question is answered, a bouquet of other problems has usually blossomed.

Natural scientists have progressed in inventing hypotheses about the sources of observed reductions in forest growth. In spite of abundant evidence in the psychology literature that individuals suffer sharp utility losses with rather minor reductions in conditions of standing stocks, economists have rarely attempted to grasp

[5]One hypothesis is nested in another when the latter implies or logically contains the former.

the value implications of these reductions.[6] As ICF Incorporated (1983, p. 10) rightly pointed out, the Calish et al. (1978) figure of $77 per acre annually that Crocker (1980, 1983) used for the value of standing forest vegetation "... is highly questionable". Two recent empirical but, as yet, unpublished studies done in Colorado and California perhaps make the figure slightly less questionable. In a study of southern California's San Bernardino's National Forest, Crocker (forthcoming) found that existing hikers and campers valued mature timber stands that had not been harmed by ambient oxidants at as much as $88 per acre annually. Walsh et al. (1981) studied 1.2 million acres of designated wilderness in Colorado, finding that an average acre had an annual value of $81. Both of these studies employed contingent valuation survey techniques, and the aggregation procedures they used to arrive at the per acre valuations required some rather stringent assumptions. The fact nevertheless remains that no evidence has yet been brought forth which contradicts the $77 per acre per year that Crocker (1980, 1983) adopted for valuing standing stocks of forest vegetation, and that the two studies in the West which have been done offer some small degree of positive support for the adopted figure.

Direct Agricultural Effects

With estimated losses amounting to about $1 billion annually, direct agricultural damages were assigned third place in the Crocker (1980, 1983) ranking. At the time the estimate was made, fears that acid deposition would directly harm agricultural plants by causing foliar necrosis, the stripping away of pesticides, reductions of leaf areas, leaching of leaf surface minerals, and cuticular erosion were prominent in the plant science literature, e.g., Cowling (1978). The quantitative dose-response information to support these fears was limited to a few studies on commercially minor crops. However, Toepfer et al. (1980) estimated that because of acid deposition, U.S. output of soybeans, tobacco, potatoes, and tomatoes could decline by 2 to 3 percent in the next half century. These apprehensions made acid deposition impacts plausibly appear as serious as the much better known direct impacts of ambient ozone upon agriculture. Nonetheless, in 1980, a draft version of Adams, et al. (1982) was the only study of the impacts of air pollution upon agricultural output and product value which captured price responses and agent adaptations. It reported that the ambient ozone levels prevailing in southern California in the mid-1970's had reduced by 3 percent the farm-gate values of the local outputs of 14 annual crops. Given the prevailing opinion in the literature that acid deposition impacts could readily prove to be as significant in yield terms as ambient ozone, and further given this author's strongly held view that price effects and agent adaptations very frequently explain more about observed outputs than do the dose-response relations that triggered them, a direct transfer of the southern California ozone results to acid deposition in the East did not seem uninformed, if the purpose of the exercise was to provide a "best guess."

In retrospect, the Crocker (1980, 1983) estimate of $1 billion was probably wrong. During the last 5 years, the plant science literature has evolved from discussions of alternative damage processes, conditional upon the existence of

[6]The primary efforts in economic theory are Berck (1981), Hartman (1976), and several working papers by Michael D. Bowes at Resources for the Future, Inc.
For examples of the relevant psychology literature see Daniel et al. (1979), and Buyhoff and Leuschner (1978).

damages, to debates about the existence of damages. Statements such as those of the Office of Technology Assessment (1984, p. 224) that the "evidence is mixed," "no general trend appears," and similar positions are now frequent. No longer does the discussion make it appear that the harmful direct agricultural impacts of acid deposition could readily rival those of ambient ozone. In fact, all the plant science research sponsored by the National Acid Precipitation Program has found only one major crop, soybeans, for which yields are affected negatively by acid deposition. Some recent preliminary work by R.M. Adams and B.A. McCarl which does account for price effects and some agent adaptations indicates that the original $1 billion estimate exaggerated control benefits for agriculture by a factor of 2 or more.[7]

Aquatic Ecosystem Effects

In spite of the research and the mass media attention that the aquatic ecosystem impacts of acid deposition got and continue to get, Crocker (1980, 1983) concluded that they were insignificant relative to materials and terrestrial ecosystem effects. In 1979, some 200 lakes and ponds in the Adirondacks were the only places in the U.S. where strong associations between increases in acidity and declines in fish populations had actually been documented. Moreover, with but a few minor exceptions, knowledge was diffuse about which other fresh-water bodies had experienced accelerated acidification or were susceptible to accelerated acidification from atmospheric pollution inputs. Even if the number of lakes, ponds, and streams in the eastern U.S. displaying a positive association between acid deposition inputs and loss of fish populations had double or tripled, the economic consequences (relative to estimated materials and terrestrial effects) would have been slight. Many thousand alternative fresh-water fishing opportunities were available. In northern New England, northern New York, and the Lake States, a dozen or more are usually available within a very few miles of any single site. Of the several classes of acid deposition impacts for which estimates of economic consequences were provided in Crocker (1980, 1983), the estimate for aquatic ecosystems was thought by the author to be the most secure.

The security of the Crocker (1980, 1983) estimate has been somewhat withered by natural science findings in the last 5 years about the extent of and the potential for fresh-water acidification that is usually harmful to fish and other aquatic organisms. According to the Office of Technology Assessment (1984, pp. 81-82), there are nearly 400 additional lakes in the Northeast that direct water chemistry measures have shown to have experienced "significant" increases in acidity. In an appendix, the Office of Technology Assessment (1984) employs a model which generates an estimate that, by 1980, 18 percent of the lake acreage and 19 percent of the stream miles in the eastern U.S. had recently been exposed to acid inputs sufficient in quantity to cause their "... aquatic resources to be highly sensitive to even low levels of additional acidic inputs" (p. 208). Nearly 3,000 lakes and 23,000 stream miles were estimated to have recently reached this status. In the Northeast, 31 percent of the lakes and 25 percent of the stream miles were said to have sunk to this state. Some relatively small amount of acid-mine drainage undoubtedly registers in these estimates; nonetheless, impacts of these magnitudes, given that they also correspond approximately to reductions in fish and waterfowl populations, do affect substantially the substitution

[7]Telephone conversation of the author with R.M. Adams, September 21, 1984. Professor Adams' work was done under contract with the National Acid Precipitation Assessment Program.

possibilities available to outdoor recreationists. The Crocker (1980, 1983) estimate of $250 million in acid deposition induced losses to aquatic ecosystems seems overly conservative when viewed from the perspective of the information available in 1984.[8] Given that Russell and Vaughn (1982) have recently estimated $0.5 to $1 billion in benefits for an increase of only 2.1 percent of "fishable" water acreage in the lower 48 states, at least a doubling or tripling of the original estimate seems safe.

Other Effects

The economic value consequences of acid deposition effects upon human health, water supply systems, and agricultural soils were also said in Crocker (1980, 1983) to be small. It was noted that the increased levels of acidity in agricultural soils caused by applications of inorganic fertilizers are commonly countered by periodic soil amendments of ground limestone. No evidence could be found that the reductions in soil acidity and/or reduced liming caused by less acid deposition would result in more than a few million dollars in control benefits. Evidence that would counteract this view has not been forthcoming.

At first glance, the economic consequences of water-supply acidification could be considerable. Although the Safe Drinking Water Committee (1977) does not adopt an unequivocal position, it does state that evidence for soft water being a causal agent in cardiovascular disease "... is sufficiently compelling so that 'the water story' is plausible...." Small disease incidences due to the inorganic and organic solutes and the microbiological agents whose human health degrading potentials are activated by low pH levels can result in large economic losses. For example, about half of the 2-million annual deaths in the U.S. are attributed to cardiovascular diseases. The Safe Drinking Water Committee (1977) estimates that these deaths could be reduced by as much as 15 percent with "optimal conditioning" of drinking water. Insofar as any lack of water conditioning is due to acid deposition, however, it is readily countered by the addition of lime costing only a few cents per million gallons. In 1970, according to the USDA's Economic Research Service (1974), about 30 billion gallons of water were daily withdrawn in the entire U.S. for rural and domestic uses.

Similarly inexpensive liming procedures are available for household, commercial, and industrial water supplies that will not be used for internal human consumption. However, in instances where the capital facilities for treatment are not already in place, annualized treatment costs could be substantial. A number of sources such as Barton (1978) and Bituminous Coal Research, Inc. (1971) indicate that the 1978 cost, including amortization of capital facilities, of raising 4.0 pH industrial water to 7.5 pH would be about 85 cents per 1,000 gallons. The

[8]Large-scale liming of either forest or aquatic ecosystems is probably technically infeasible. In agricultural soils, lime is mixed with the soil. It is not evident how this could be done in a forest. As Holden (1979) emphasized, the effective use of lime in aquatic systems requires a great deal of information about the hydrological and chemical properties of each water body. One might reasonably question whether enough limnologists and water chemists could be produced to generate and update this knowledge. If small-scale liming is occasionally feasible, the economic losses imposed by acidification would then be the costs of restoration by liming, plus the differences between the value of the aquatic activities missed and the value of the substitute activities adopted after acidificiation and prior to restoration.

implications of this for increased water treatment costs must remain unknown until information on the spatial and temporal distributions of acidified water supplies, acceptable pH levels, and total water usages is aquired.

ERRORS OF COMMISSION AND OMISSION

Adaptations and Price Effects

If one is to accept the estimates and revisions of the previous section, he must accept the author's method as well his judgments. For reasons already stated, the method, which simply involved the multiplication of invariant prices by others' estimates or conjectures of reduced service flows from existing stocks of biological and physical assets, is incomplete, perhaps seriously so. Since a reduction in acid deposition will relax a presumedly binding constraint that economic agents now confront, thereby expanding the set of alternatives from which they can choose, the failure to consider adaptation opportunities imparts a downward bias to the estimated benefits of reduced acid deposition. The agent might discover a production or consumption pattern under the new and lower pollution regime that makes him better off under that regime than would the old pattern. Moreover, if prices are not allowed to vary, consumers can have nothing to gain from acid deposition reductions which increase supplies of the goods they consume. If price variations had been allowed, the producers of these goods could be either winners or losers.[9]

Given that one accepts the natural science estimates and conjectures on service flow effects that were adopted in the previous section, and further given that the author's selective judgments are reasonable, the failure to account for price effects and adaptations must mean that the Crocker (1980, 1983) estimate of as much as $5 billion in gross benefits annually from preventing acid deposition is too low. This argument, however, does not exhaust the reasons why the maximum $5-billion estimate might readily be too low.

Classes of Effects not Considered Earlier

The precursors of acid deposition are widely acknowledged to be primarily sulfur dioxide and nitrogen oxide gaseous emissions from human production and consumption activities. Unless neutralizing substances such as an alkali or ammonium are present, the oxygen in the atmosphere and in moisture oxidizes these gases transforming them into sulfate and nitrate particles. When moisture is

[9]Let the producer's total revenue (TR) be px, where p is the unit price of the commodity and x is its quantity. Price is related to quantity by the inverse demand function, $p = p(x)$. Marginal revenue (MR) is defined as the change in TR with respect to a one-unit change in x, or:

$$MR = \frac{d(TR)}{dx} = p + x\frac{dp}{dx} = p(1 + \frac{dp}{dx}) = p(1 + \frac{1}{\varepsilon}).$$

Since p and x are inversely related, ε is always negative. Thus $MR > 0$, when $\varepsilon < -1$. TR thus rises when x increases, and, consequently, price falls. Alternatively, when $-1 < \varepsilon < 0$, $MR < 0$. TR therefore falls when x increases.

available, whether it be in the atmosphere, on land surfaces, or on vegetation or materials, the particles can become acidic. In short, whether the chemical transformation takes place in the atmosphere some distance removed from the source or on the ground in the immediate vicinity of the source, acidic particle formation is the result.

The preceding qualitative chemical truths about the paths by which acid deposition is formed imply that traditional public and policymaker discourse about "acid rain" and long-distance transport tends to obscure some classes of acid deposition effects, causing the boundaries of the entire set of effects to be drawn too narrowly. In particular, atmospheric visibility losses and human respiratory system effects are excised from the discussion context. Neither was considered in Crocker (1980, 1983). The benefits of substantially reducing these effects are plausibly large, perhaps even larger than the estimated benefits of a major lessening of the suspected materials or forest ecosystem effects. According to the Office of Technology Assessment (1984, p. 254) sulfate particles account for about 53 percent and nitrate particles account for another 8 percent of visibility impairment in the U.S. By extrapolating from a set of contingent valuation studies of atmospheric visibility in western locations, Mathtech, Inc. (1981) estimated that a 20-percent improvement in visual range in all regions of the 48 contiguous states would generate $10-$20 billion in annual benefits. Freeman (1982, p. 128), after having carefully reviewed numerous studies, concluded that $17 billion (in 1978 dollars) was the "most reasonable point estimate" of the annual human health benefits realized from the air pollution reductions that occurred in the U.S. over the 1970-1978 period. Most of these estimated benefits were attributed to reductions in sulfate and nitrate particles or their gaseous precursors. Given these estimates, it is conceivable that the Crocker (1980, 1983) estimate of the maximum gross economic benefits of controlling the acid deposition levels of the late 1970's is too small by a factor of 2 or more.

Increasing Incremental Benefits

In addition to treating unit prices as constant, the simple multiplication method employed in Section II assumes that changes in service flows are constant across all levels of acid deposition. Unless changes in service flows are very small, the validity of assuming these flows to be constant is extremely doubtful. With constant prices and decreasing marginal improvements in service flows as acid deposition levels decline, the simple multiplication method would cause control benefits to be overestimated; increasing marginal improvements in service flows would cause them to be underestimated. Most treatises on the economics of pollution problems take the former as true, i.e., that the marginal benefits of pollution control decline. Crocker and Forster (1981, 1984) have reviewed several natural system, dose-response function examples where this standard economic premise is clearly violated for at least some range of acid deposition control levels. Their examples encompass a broad effects spectrum, including effects on soils, aquatic ecosystems, and atmospheric visibility.[10] In fact, any acid deposition dose-response function which has a damage threshold is likely to exhibit the largest incremental reductions in service flows in the immediate neighborhood of the threshold.

[10]As Violette (1984) points out, these increasing marginal service flows from individual assets do not necessarily imply that aggregate marginal benefit functions will be increasing.

Any propensity for the presence of increasing marginal benefits of acid deposition control need not reside solely in dose-response functions for the affected physical and biological assets. Indeed, people's utility or satisfaction responses to the aesthetic impacts of acid deposition upon vegetation and materials may also be responsible for increasing marginal benefits of control. Baird and Noma (1978) review numerous psychophysical experiments involving human perceptions of sensory events such as light and color changes. These perceptions have been codified in the form of Fechner's law, which states that the strength of a just noticeable increment in a sensation is proportional to the logarithm of its stimulus. If satisfaction is properly regarded as a sensation, then Fechner's law implies that the aesthetic impacts of acid deposition will have a progressively declining negative effect upon utility, whether utility be scaled in economic or psychological terms. To date, however, there appears to be only a single study that has obtained this result while employing an economic scale. Crocker (forthcoming) found that outdoor recreationists attached strongly decreasing marginal losses to oxidant-induced damges to the vegetation stock in California's San Bernardino National Forest. Since the biological damage function was linear and since endogenous variables such as visit frequency were controlled, the source of the increasing marginal benefits had to be identified with the recreationists' tastes, i.e., the distributions of their preferences over their perceptions of damages. Since the appearance to the layman of oxidant-damaged forests is quite similar to that of acid deposition damaged forests, it is not farfetched to surmise that the marginal benefits of reducing aesthetic damages to the latter forests would behave in much the same fashion.

Irreversibilities

The acid deposition literature is replete with speculations, hints, and consensus scientific statements about the mining of ecosystem nutrients and the accumulation of ecosystem toxins. Johnson (1981) provides evidence that the current and future consequences of current and past acidification are not fully reversible; the incremental benefits of control are not invariant with respect to the status-quo point and the direction of movement. This irreversibility property means that efforts to estimate the economic consequences of acid deposition effects must account for the loss of future opportunities to enjoy ecosystem amenities and life support services. Completeness requires that the foreclosed options of ecosystem users be taken into account. An analysis that considers only the value of the reduction in current visible flows will inevitably underestimate the benefits of control.

This conclusion can be reached without the fact of foreclosed options. As Arrow and Fisher (1974) show, the bare possibility that current actions might burden and constrain or deplete future opportunities must be counted as a cost against the current action. Subsequent information about technologies, price structures, tastes, or acid deposition effects can be exploited only if irreversible consequences have been avoided. Otherwise, the consequences of a decision that might result in acidification cannot be undone, even if new information suggests that the decision was regrettable.

Inability to Pool Risks

For many bundles of public decision problems, one can reasonably assume, in accordance with the risk-pooling argument of Samuelson (1964), that he who ultimately loses on one project will ultimately gain on other projects. Thus, other than accounting for the individual's realized losses or gains, no special weight need be given to the risk the individual confronts with respect to the outcome of a single project. Arrow and Lind (1970) make an analogous risk-spreading argument for single projects affecting large numbers of people: as numbers increase, the individual's exposure to overall project risk declines. However, Fisher (1973) has pointed out that the risks associated with many pollution impacts have "public bad" attributes; that is, the risk that the individual faces is unchanged by increases in the number of individuals who confront the same risk, implying that the individual's share of the risk premium will not decline with increases in the number of affected individuals. Society's risk and the individual's risk may be undistinguishable. This seems a more-or-less apt description of a key feature of the acid deposition problem, since the risks it poses are highly correlated across individuals. Adding some individuals in Georgia to those whose assets are at risk from acid deposition will not reduce the implicit risk premia that individuals in the State of Maine must pay. The removal of these premia must be counted as one of the benefits of controlling acid deposition.

At the level of the individual who might encounter acid deposition effects, further support is available in the economics and in the psychology literature for awarding risk premia. In terms of the distinction in the introduction to this paper between citizen beliefs that may or may not be worthy of policymaker attention and citizen tastes which the policymaker is expected to heed, these findings refer to tastes. For example, Fischoff et al. (1978) have found that individuals find a given degree of risk to be less bearable if they are unable to influence it, if their exposure to it is involuntary, or if it has potential delayed effects. These three properties certainly describe features of the situations that challenge owners of assets exposed to acid deposition. All are costly in their own right.[11]

A BAYESIAN PERSPECTIVE

It is an unfortunate fact of human existence in general and the acid deposition problem in particular that practical decisions must be made, and that beliefs about the consequences of these decisions must be formulated with incomplete information. Any claim that the natural science and economic foundations of the discussions in previous sections cover all feasible world states relevant to the economic benefits of acid deposition control would be preposterous. Only the provisional claim can be made that those world states which the discussion neglects will not have enough influence over ultimate economic consequences to falsify the predictions. Falsification requires contrary and broadly accepted evidence, which, as yet, has not been, but ultimately could be, advanced.

Nonetheless, nearly everyone would assign a prior probability less than unity but greater than zero to the proposition that acid deposition is harmful enough to

[11]Involuntary risks remove the ex ante opportunity to adapt, lack of influence removes both ex ante and ex post opportunities to adapt, and delayed effects cause subsequent as well as present decisions to be risky. Spence and Zeckhauser (1972) show that each of these properties imposes economic costs. See also Munera and de Neufville (1983) for a review of models in which uncertainty is an argument in the individual's objective function, i.e., where individuals have a taste for the process by which a particular choice problem is resolved.

cause the annual gross benefits in 1978 dollars of its control to be as much as $5 billion. However, given available information and personal predispositions, the specific prior assigned probability would differ widely across individuals. Given the observations that I have thus far synthesized, I feel (my prior is ...) the chances to be very high that acid deposition is sufficiently harmful to cause the proposition to be true. The previous discussion is intended to convince those with alternative inclinations that this prior belief has substantial truth value. Considering, then, the wide degrees of prior belief in the credibility of the hypothesis, is there a set of criteria, as Savage (1954) originally asked, by which these prior beliefs may be evaluated; that is, can one distinguish among prior judgments on the basis of whether or not they make best use of existing observations about the effects of acid deposition?

In order to grasp the significance of this question, it is necessary to appreciate the roles that prior beliefs play in interpretations of existing observations, which are the sample evidence. If one views the world according to Bayes (1764), the ultimate degree of belief (the posterior) accorded a hypothesis is a weighted combination, as Zellner (1971) shows, of prior beliefs and sample observations. In particular, let T^+ be a set of observations consistent with the presence of an acid deposition effect of given magnitude, and let $T-$ be the opposite. Similarly, define H^+ as the hypothesis which posits the effect, and let $H-$ be the negation of this hypothesis. Assorted conditional probabilities can be denoted as $p(T^i|H^j)$, where $(i,j = +,-)$, and p is a probability value. For example, $p(H^+|T^+)$ is the probability of the hypothesis being true, given that an observation consistent with the hypothesis has been made. One's knowledge of the frequency of each combination of observation and hypothesis is summarized by Bayes' theorem,

$$p(H^+|T^+) = \frac{p(H^+)\cdot p(T^+|H^+)}{p(H^+)\cdot p(T^+|H^+) + p(H^-)\cdot p(T^+|H^-)}$$

which states the probability that the hypothesis which posits the effect is correct when a new set of observations consistent with the hypothesis is made.

So as to clarify the importance of prior beliefs, and the logical demands that observations consistent with the hypothesis of harmful acid deposition effects make upon these beliefs, consider the policymaker who thinks that there is no more than a 2-percent chance ($p(H^+) = .02$) that acid deposition control would generate as much as $5 billion in annual benefits. Suppose that new natural science data on effects is brought to his attention which is consistent with the hypothesis that the annual benefits of controlling acid deposition are $5 billion. This new data might refer, for example, to the extent of freshwater acidification and its impact on aquatic ecosystems. The policymaker's decision problem has been altered. Suppose he had earlier been told by an economist, whose diagnostic skills he trusts, that there is only a 10-percent chance that the $5-billion control benefits hypothesis would ultimately be observed,[12] i.e., that $p(T^+|H^+) = .10$. Further assume that it is known from the natural science literature that the newly observed effects would occur only about 1 percent of the time in the absence of acid deposition, i.e., that $p(T^+|H^-) = .01$.

[12]The discrepancy between the judgments of the policymaker and the economist might arise because the economist has devoted more time to uncovering the implications of alternative world states and/or because the economist looks at the world more narrowly.

According to Bayes' theorem, the policymaker's updated or posterior probability, $p(H^+|T^+)$, of the truth of the hypothesis must then be:

$$p(H^+|T^+) = \frac{(.02)(.10)}{(.02)(.10) + (1.00-.02)(.01)} = \frac{.0020}{.0020 + .0098} = .169.$$

In short, if the policymaker employs the new evidence in a coherent fashion, he must revise his opinion from 2 percent to 17 percent that the benefits of controlling acid deposition will be as much as $5 billion annually. If his initial opinion about the chances of the hypothesis being true were 5 rather than 2 percent, the new evidence would require a similar opinion revision to 34 percent. In short, even if one gives very small initial credence to the proposition that acid deposition effects have substantial economic consequences, a bit of evidence that supports the proposition can necessitate a major revision in beliefs. Alternatively, contrary evidence would cause a substantial downward revision in a strong prior that held the hypothesis to be true. Generally, the less the degree of prior belief, the higher in relative terms will be the posterior probability when new evidence consistent with the hypothesis emerges. Moreover, the gambles the policymaker is willing to undertake in the light of this new evidence will vary widely with his priors. His judgments depend not only on the quality of the information available to him, but also on how he uses this information to formulate his prior views on the consequences that will ultimately be realized.

As repeatedly noted earlier, given the available numerical and qualitative natural science and economic information, my prior is high that annual acid deposition control benefits are as much as $5 billion. Is this prior well calibrated? It is hard to say precisely, but some general criteria can be set forth.

In the absence of complete knowledge, some arbitrary choices among partially coherent numerical and qualitative observations are unavoidable. Of course, the more that is known, the fewer the arbitrary choices that must be made. If relatively little is known, the trick is to make the least arbitrary, coherent set of choices [Harsanyi (1976)].

As is well-known, there is a tradeoff between model elaboration and tolerance to measurement error. Generally, the greater the number of attributes introduced into a model in the form of properties that must be directly measured, the more likely are some pairs of these properties to be highly correlated, and therefore, the less the degree of precision in any particular estimate. Consider, for example, the hypothesis that the economic consequences of acid deposition induced reductions in commercial timber yields are a function only of yield reductions from current standing stocks and existing stumpage prices. Compare this to a hypothesis which introduces a number of additional parameters, such as the prices of timber substitutes, forest successional patterns, and expectations about future forest management technologies. The more complex hypothesis does not predict anything about these latter parameters; it requires that they be directly observed. If observation is impossible or extremely costly, use of the more complex hypothesis to construct one's estimates necessitates the arbitrary assignment of specific numerical values to these additional parameters. As Klepper and Leamer (1984) have demonstrated, imprecise measures of nonorthogonal secondary influences such as forest successional patterns "smear" the influence of primary variables such as stumpage prices. Thus, although the assignment of zero values to these secondary influences is no less arbitrary than assigning them some nonzero value, by excluding them from the exercise, one avoids confounding their measurement errors with the influence of more relevant and precisely measured parameters.

The preceding argument can be made in a slightly different and more succinct fashion. Because the assignment of values to the secondary parameters is more-or-less arbitrary, any particular assignment to a parameter is only one of a near-infinity of alternative assignments, none of which have any structured knowledge to recommend them. Thus the prior probability weight given to a particular assignment must be infinitesimally small. Accordingly, any specific version of a hypothesis which includes these parameters must be assigned a prior probability much smaller than a less complex hypothesis which includes only the well-measured parameters. It certainly can be argued as to whether or not the parameters used to arrive at my strong $5 billion prior are primary or secondary; however, with perhaps a few exceptions, the parameters that were used were those with the most precise available measurements. Because of the incompleteness of broadly accepted knowledge about the natural and economic factors that influence the economic consequences of acid deposition control, individual (generally lower) estimates derived from specific versions of hypotheses more complex than Crocker (1980, 1983) must usually be assigned a substantially smaller prior probability. However, as the elementary application here of Bayes' theorem demonstrated, any lessening in arbitrariness of value assignments to these secondary parameters could readily oblige a major revision in the strong prior reviewed in this paper.

CONCLUSIONS

As are nearly all priors, the correctness of the prior set forth here is problematic. What matters at this very limited stage of our understanding of the economic consequences of acid deposition control is that the problematic be taken seriously. Hypotheses can be more finely tuned (made more complex), because by no means have all the parameters been studied (e.g., the prices of substitutes) for which modern economic assessment techniques allow quite exact measures to be obtained.[13] Very little is to be gained, however, by inventing stories involving parameters for which no realistic investment of scarce research resources is likely to provide an acceptably exact measure of value or form. Their truth value is very small, they dissipate research energies, and they probably contribute to a paralysis of political will.

REFERENCES

Adams, R.M., Crocker, T.D., and Thanavibulchai, N., 1982, "An Economic Assessment of Air Pollution Damages to Selected Crops in Southern California," *J. Environ. Econ. Manage.*, 9:42.

Arrow, K.J., and Fisher, A.C., 1974, "Environmental Preservation, Uncertainty, and Irreversibility," *Quart. J. Econ.*, 88:302.

[13]Freeman (1979) provides an able and reasonably up-to-date survey of these techniques.

Arrow, K.J., and Lind, R.C., 1970, "Uncertainty and the Evaluation of Public Investment Decisions," *Amer. Econ. Rev.*, 60:364.

Baird, J.C., and Noma, E., 1978, "Fundamentals of Scaling and Psychophysics," John Wiley and Sons, NY.

Barton, P., 1978, "The Acid Mine Drainage," in: *Sulfur in the Environment: Part II*, J.O. Nraigu, ed., John Wiley and Sons, NY.

Bayes, T., 1764, "An Essay Towards Solving a Problem in the Doctrine of Chances," *Philosophical Transactions Giving Some Account of the Present Undertakings, Studies, and Labours of the Ingenious in Many Considerable Parts of the World for the Year 1763*, Vol. 53, The Royal Society, London, England.

Berck, P., 1981, "Optimal Management of Renewable Resources with Growing Demand and Stock Externalities," *J. Environ. Econ. Manage*, 8:105.

Bituminous Coal Research, Inc., 1971, "Studies of Limestone Treatment of Acid Mine Drainage, Part II," USEPA, Washington, DC.

Buyhoff, G.J., and Leuschner, W.A., 1978, "Estimating Psychological Disutility from Damaged Forest Stands," *For. Sci.*, 24:425.

Calish, S., Fight, R.D. and Teeguarden, D.E., 1978, "How Do Nontimber Values Affect Douglas-Fir Rotatations?", *J. of For.*, 76:217.

Commission on Natural Resources, National Academy of Sciences, 1975"Air Quality and Stationary Source Emission Control," Committee on Public Works, U.S. Senate, 94th Cong., 1st Sess.

Cowling, E.B., 1978, "Effects of Acid Precipitation and Atmospheric Deposition on Terrestrial Vegetation," in: *A National Program for Assessing the Problem of Atmospheric Deposition (Acid Rain)*, J.N. Galloway et al., eds., Natural Resource Ecology Laboratory, Colorado State Univ., Fort Collins, CO.

Crocker, T.D., "On the Value of the Condition of a Forest Stock," *Land Econ*, (forthcoming).

Crocker, T.D., 1982, "Pollution Damage to Managed Ecosystems: Economic Assessments," in: *Effects of Air Pollution on Farm Commodities*, J.S. Jacobson and A.A. Millen, eds., The Izaak Walton League of America, Arlington, VA.

Crocker, T.D., 1984, "Scientific Truths and Policy Truths in Acid Deposition Research," in: *Economic Perspectives on Acid Deposition Control*, T.D. Crocker, ed., Butterworth Publishers, Boston, MA.

Crocker, T.D., 1980, Statement, "Economic Impact of Acid Rain," U.S. Senate, 96th Cong., 2nd Sess.

Crocker, T.D., 1983, "What Economics Can Currently Say About the Benefits of Soil Deposition Control," in: *Adjusting to Regulatory, Pricing, and Marketing Realities*, H.M. Trebing, ed., Institute of Public Utilities, Michigan State Univ., East Lansing, MI.

Crocker, T.D., and Forster, B.A., 1984, Author's reply, *J. Air Poll. Cont. Assoc.*, 34:42.

Crocker, T.D., and Forster, B.A., 1981, "Decision Problems in the Control of Acid Precipitation: Nonconvexities and Irreversibilities," *J. Air Poll. Cont. Assoc.*, 31:31.

Daniel, T.C., Zube, E.M., and Driver, B.L., eds., 1979, "Assessing Amenity Resource Values," Rocky Mountain Forest and Range Experiment Station, Fort Collins, CO.

Economic Research Services, U.S. Dept. of Agriculture, 1974, "Our Land and Water Resources," USGPO, Wash., DC.

Fischoff, B., Slovic, P., Lichtenstein, S., Read, R., and Combs, R., 1978, "How Safe is Safe Enough? A Psychometric Study of Attitudes Toward Technological Risks and Benefits," *Pol. Sci.*, 9:127.

Fisher, A.C., 1973, "A Paradox in the Theory of Public Investment," *J. Public Econ.*, 2:321.

Freeman, A.M. III, 1982, "Air and Water Pollution Control," John Wiley and Sons, NY.

Freeman, A.M. III, 1979, "The Benefits of Environmental Improvement," The Johns Hopkins Univ. Press, Baltimore, MD.

Greenfield, S.M., Simmons, W.S., and Seigneur, C., 1983, "A Reexamination of Suggested Benefits that Might be Associated with a Reduction of SO_2 Emissions from Consumers Power Company's Cobb and Campbell Fossil Fuel Plants," Systems Applications, Inc., San Rafael, CA.

Haagenrud, S.C., and Atteraas, J., 1980, "Atmospheric Corrosion Testing," Hollywood, FL.

Harsanyi, J.C., 1976, "Essays on Ethics, Social Behavior, and Scientific Explanation," D. Reidel Publishing Co., Boston, MA.

Hartman, R., 1976, "The Harvesting Decision When a Standing Forest Has Value," *Econ. Inq.*, 4:52.

Haynie, F.H., Spence, J.W., and Upham, J.B., 1976, "Effects of Gaseous Pollutants on Materials--A Chamber Study," USEPA, Wash., DC.

Hicks, J.R., 1939, " The Foundations of Welfare Economics," *Econ. J.*, 49:696.

Holden, A.V., 1978, "Surface Waters," in: *Ecological Effects of Acid Precipitation*, M.J. Wood, ed., Central Electricity Research Laboratories, Surrey, U.K.

ICF Incorporated, 1983, "A Review and Critique of the Crocker, Tschirhart, and Adams Assessment of the Benefits of Controlling Acid Precipitation," mimeo.

Jacobson, J.S., 1983, "Ecological Effects of Acid Rain," *J. Air Poll. Cont. Assoc.*, 33:1031.

Johnson, A.H., 1983, "Red Spruce Decline in the Northeastern U.S.: Hypotheses Regarding the Role of Acid Rain," *J. Air Poll. Cont. Assoc.*, 33:1049.

Johnson, D.W., 1981, "Acid Rain and Forest Productivity," Environmental Sciences Division, Oak Ridge National Laboratories, Oak Ridge, TN.

Jonsson, B., 1976, "Soil Acidification by Atmospheric Pollution and Forest Growth," Proceedings of the First International Symposium on Acid Precipitation and the Forest Ecosystem, Northeastern Forest Experiment Station, Upper Darby, PA.

Klepper, S., and Leamer, E.C., 1984, "Consistent Sets of Estimates for Regressions with Errors in All Variables," *Econometrica*, 52:163.

March, J.G., 1978, "Bounded Rationality, Ambiguity, and the Engineering of Choice," *Bell J. Econ.*, 9:587.

Mathtech, Inc., 1981, "Benefits Analysis of Alternative National Ambient Air Quality Standards for Particulate Matter, Vol. II," Mathtech, Inc., Princeton, NJ.

McCarthy, E.F., Stankunas, A.R., and Yocum, J.E., 1981, "Benefit Model for Pollution Effects on Materials," TRC Environmental Consultants, Inc., Weathersfield, CN.

Munera, H.A., and de Neufville, R., 1983, "A Decision Analysis Model When the Substitution Principle is Not Acceptable," in: *Foundations of Risk and Utility Theory with Applications*, B.P. Stigum and F. Wenstop, eds., D. Reidel Publishing Co., Boston, MA.

Office of Technology Assessment, U.S. Congress, 1984, "Acid Rain and Transported Air Pollutants: Implications for Public Policy," USGPO, Wash., DC.

Panel on Nitrates, 1978, "Nitrates: An Environmental Assessment," National Academy of Sciences-National Research Council, Wash., DC.

Russell, C.S., and Vaughn, W.J., 1982, "The National Recreational Fishing Benefits of Water Pollution Control," *J. of Environ. Econ. and Manage.*, 9:328.

Safe Drinking Water Committee, 1977, "Drinking Water and Health," National Academy of Sciences-National Research Council Wash., DC.

Samuelson, P.A., 1964, Discussion, *Amer. Econ. Rev.*, 54:422.

Savage, L.J., 1954, "The Foundations of Statistics," John Wiley and Sons, Inc., NY.

Smith, V.K., 1981, "CO_2, Climate, and Statistical Inference: A Note on Asking the Right Questions," *J. of Environ. Econ. and Manage.*, 8:391.

Spence, M., and Zeckhauser, R., 1972, "The Effect of the Timing of Consumption Decisions and the Resolution of Lotteries on the Choice of Lotteries," *Econometrica*, 40:401.

Stankunas, A.R., Unites, D.F., and McCarthy, E.F., 1983, "Air Pollution Damage to Man-Made Materials: Physical and Economic Estimates," Electric Power Research Institute, Palo Alto, CA.

Toepfer, F., Boberschmidt, L., Smith, B., Keitz, E., Wisniewski, J., Pitter, R., and Kobele, A., 1980, "Acid Rain Impacts on Fish Populations and Sensitive Crops," The MITRE Corp., McLean, VA.

Tomlinson, G.H., 1983, "Air Pollutants and Forest Decline," *Environ. Sci. and Tech.*, 17246A.

U.S. Department of Commerce, 1980, "Statistical Abstract of the U.S., 1979," USGPO, Wash., DC.

United States Forest Service, 1978, "Forest Statistics of the U.S., 1977--Review Draft," USGPO, Wash., DC.

Violette, D.M., 1984, "Comment on 'Decision Problems in the Control of Acid Precipitation: Nonconvexities and Irreversibilities,'" *J. Air Poll. Cont. Assoc.*, 34:42.

Walsh, R.G., Gillman, R.A., and Loomis, J.B., 1981, "Wilderness Resource Economics: Recreation Use and Preservation Values," Colorado State Univ., Fort Collins, CO.

Zellner, A., 1971, "An Introduction to Bayesian Inference in Econometrics," John Wiley and Sons, Inc., NY.

RESPONSE TO THOMAS D. CROCKER

A. Myrick Freeman III

Professor of Economics
Bowdoin College
Brunswick, Maine 04011

My comments will be organized around three major points. The first is my response to the charge to discuss areas of disagreement and controversy in the assessment of benefits. The second is a discussion of the usefulness of and limitations to the quantitative estimates of damages provided in Tom Crocker's paper. And the third is a plea to adopt a more comprehensive perspective on the problem of long-range transport of certain atmospheric pollutants.

Respondents were asked to present alternative viewpoints and to discuss areas of controversy within the realm of the topics covered by the papers assigned to us. I find this a hard charge to respond to, because I think that there is substantial agreement within the economics profession on most of the major points made either explicitly or implicitly in Crocker's paper. Let me take a moment to outline these major points of agreement so that anyone who wants to take exception to them can do so.

First, there is general agreement among economists on the proper way of defining benefits and damages. As outlined in Crocker's paper, this is in terms of individuals' preferences and is measured by compensating income changes. Furthermore, there is general agreement on the conceptually correct economic models for organizing and analyzing data to estimate the monetary damages of various kinds of effects. For example, in estimating effects on commercial forestry, it is necessary to model supply and demand in order to capture producer and consumer responses to both changes in physical yield and the resulting changes in prices. For materials damage, we must know the range of substitution and mitigating possibilities as well as the effects on cost and price in order to estimate damages or benefits that correspond to the theoretical definitions.

Before moving on to other points of agreement, I must note that there are two areas of continuing controversy concerning concepts and methods for estimating acid deposition control benefits. The first has to do with the appropriate way of defining and the empirical significance of a variety of so-called intrinsic or nonuse values. These include such things as option value, bequest value, and preservation and existence value; and they refer to individuals' willingness to pay to prevent irreversible changes to environmental resources which might result in the extinction of certain species or the destruction of unique ecological systems. Such intrinsic and nonuse values might be very important in connection with changes to terrestrial and aquatic ecosystems as a consequence of acid deposition.

The second area of controversy has to do with the accuracy and reliability of the contingent valuation method of obtaining estimates of monetary benefits and damages. The models I referred to earlier on which there is general agreement about validity all have the property of relying on observations of actual behavior in response to changes in pollution levels, costs, and prices. In the contingent valuation method, people are asked to assign dollar values on the basis of "what if" questions. The contingent valuation method has both the virtue and the defect of being based on responses to hypothetical rather than real changes. This is a virtue in that the contingent valuation method can be used where it is too costly or difficult to obtain data on actual behavior or where the environmental changes of interest have not actually occurred. The defect arises from our uncertainty concerning the degree of correspondence between answers to hypothetical questions and actual responses to real changes. This controversy is relevant to Crocker's paper since the contingent valuation method was the basis for evidence he cited concerning the lost recreational value due to degraded forest ecosystems and the damages due to reduced atmospheric visibility.

The second area of agreement is that the conceptually correct models often place severe demands on the economic data, especially for pollution problems involving large areas of the country or large displacements from the unpolluted state. The conceptually correct model may be complex because of the necessity to capture a variety of behavioral responses to the pollution. But increasing model complexity is costly in terms of gathering the necessary data, in terms of analysis, and in terms of loss of accuracy, as Crocker points out. This leads to the third area of agreement.

Damage and benefit estimates based on much simpler models can still be very useful for policymaking. These models typically involve multiplying a predicted change in quantity by an assumed unchanged price (for example as in Crocker's forestry damage) or multiplying a predicted change in price or cost by an assumed unchanged quantity (for example as in Crocker's materials damage). These simpler models are more accurate the smaller is the range of substitution possibilities and responsiveness of producers and consumers to price or cost changes. Also in many cases it is possible to predict the direction of any bias involved in using a simple model. For example, if the benefits of a pollution control program were estimated by multiplying a predicted decrease in production costs by an assumed unchanged quantity being produced, this would yield an underestimate of the true benefits. This is because the simple multiplication does not capture the additional benefits due to the increase in supply of the good in response to the lower cost.

The discussion so far has focused on the economic dimension of benefit estimation models. There is general agreement that, especially in the case of acid deposition, the physical and natural science foundation on which economic models must rest is not adequate. Estimates of the benefit to commercial forests of an X-percent reduction in sulfur emissions require predictions of the change in ambient sulfate concentrations and deposition in the affected forest area as well as predictions of the change in tree growth that results from cleaner air and less acid deposition. Economic benefit and damage estimates must therefore include phrases such as "if what the scientists tell us about the effects of acid deposition on tree growth is true," or "assuming that the relationship is such and such ..."

There has been relatively little sophisticated and rigorous economic analysis of the benefits of controlling acid deposition. I suspect that one reason for this is that economists have recognized that it would be wasteful of scarce analytical resources to develop rigorous economic models that have inadequate scientific foundations. I predict that, as the basic science becomes better understood, the rigor and sophistication of the economic models will grow rapidly.

An example of this pattern of economic analysis following basic scientific developments can be found in the recent efforts of EPA to estimate the benefits of controlling ozone pollution. The National Crop Loss Assessment Network experiments have provided estimates of dose-response functions for various crops under field conditions at ambient pollutant concentrations. These dose-response functions have become the basis of an economic model of the national agricultural economy reflecting substitution patterns among crops by farmers and price and quantity effects in response to yield changes. These models are far more defensible on conceptual or theoretical grounds than earlier models for estimating the agricultural benefits of air pollution control. And incidentally they are producing estimates of potential control benefits that tend to be substantially higher than those produced by the earlier less sophisticated models.

The second general area I want to discuss is the usefulness of Crocker's damage estimates for informing the current policy debate. What Crocker has provided is a first-cut estimate of the damages due to the emissions of precursors of acid deposition, or what is the same thing, the benefits of eliminating all such emissions. Crocker's data are not adequate for estimating the benefits of alternative feasible control schemes, that is, those involving less than 100-percent control. Often in policy analysis it is assumed that the benefit or damage function is linear, so that an X-percent control policy is predicted to yield benefits equal to X-percent of the total damages. But there is reason to believe the damage function is nonlinear in the case at hand; and as Crocker argues, it may even be nonconvex.

This is not to say that his estimates have no value. First, he has established the plausibility of the assertion that the economic benefits of a control program may be significant, assuming that what he has learned from the scientists about the physical relationships involved is true. And second, by providing estimates of damages by category, Crocker has provided a basis for allocating scarce research resources. For example, a research program which would reduce the range of uncertainty surrounding estimates of materials damages or terrestrial ecosystems damages is likely to have more value to policymakers than a program which would have a similar effect in reducing the uncertainty of the aquatic ecosystem damage estimate.

The last point I would like to discuss also is based on an examination of Crocker's estimates of damages by category, and his discussion of "errors of commission and omission." What is becoming clear is that it is at best misleading to speak of an "acid rain problem" or even of an "acid deposition problem". Rather we have a problem of long-range transport of emissions of sulfur and perhaps other pollutants to the Northeast where, perhaps in combination with local emissions, a variety of adverse consequences are felt both because of elevated ambient concentrations before deposition and because of deposition in dry or wet form of sulfate and other products.

To put it differently, the problem is not just one of a few dead fish in some small lakes up north, as at least one political figure has suggested. Crocker's largest category of damages is to materials. And most of the materials are in the urban areas of the Northeast. The materials damages due to sulfates arise from wet or dry deposition. But the potentially large categories of damages to human health and visibility arise not from deposition but from elevated ambient concentrations of these substances. And again, these damages will occur where the people are, in the urban areas of the Northeast.

Controlling emissions will reduce both damages due to elevated ambient concentrations and those due to deposition. And unless Crocker's estimates are seriously in error both as to order of magnitude and relative importance of categories, a significant share of the benefits, perhaps up to a half of any control program, will be realized in the urban areas. Thus the environmentalists who point to dead trees and deformed fish in their pleas for control of acid deposition may be playing into the hands of their opponents who argue that it is unreasonable to incur such large costs just for the benefit of a few fishermen and backpackers.

RESPONSE TO THOMAS D. CROCKER

Paul W. MacAvoy

Dean and Professor of Business Administration
Graduate School of Management
and
Professor of Economics
University of Rochester
Rochester, NY 14627

Measuring the benefits from improved environmental quality is a formidable task, even under ordinary conditions. But at this time the difficulty is increased by the intense scrutiny given to new policy proposals. Reductions in federal expenditures on all programs make it necessary to subject estimates of gains from new programs to exceptional scrutiny for credibility and even accuracy. Thus the standards for evaluating estimates have increased; no longer can the advocate wave at gains in the "quality of life" without being subject to searching examination as to whether such "quality" is more or less than that achieved from competing medical care reimbursement plans, or secondary education loan guarantees, or expenditures on highway reconstruction. Benefits that are not measurable, in the sense that they cannot be re-estimated, might as well not exist, given that in the contest for federal program outlays that are movable from one program to another there are other programs that can show measurable and significant results.

It is within this context that Professor Crocker's estimate of benefits has to be considered. He states that there are $5-billion worth of annual gains in environmental quality from acid rain reduction. The question is whether that estimate exists, in this sense of being of commensurate worth with estimates of gains from alternative programs. More practically, it is fairly well documented that to achieve significant benefits requires the reduction of sulfur emissions by eleven million tons per year, which in turn requires an annual capital outlay including interest of between $5 and $7 billion in the fifteen states on the eastern seaboard of the United States. Is Professor Crocker's estimate sufficiently precise to allow us to conclude that annual benefits are at the low end of the range of likely costs of acid rain reduction? Or is the estimate sufficiently imprecise that real-world benefits could very well either fall short of or substantially exceed the $5 to $7 billion of outlay required to reduce sulfur emissions?

In fact, the $5-billion benefit estimate has been with us since 1980, as revealed to the public in a publication from a National Science Foundation project and again in Professor Crocker's testimony before a House committee hearing on the benefits and costs of acid rain. Since it has been on the record for close to five years, it should "exist" in at least two forums if it is to be operational for decisionmaking. The first forum is the community of scholars in environmental science, public policy, and economics. Operating through the scholarly journals, conferences, and meetings, that forum produces a paper trail of comments that subject any such number to intense scrutiny.

Unfortunately, there is no record of discussion and appraisal of Professor Crocker's estimate. The index of citations in scientific publications shows an enviable number of citations to Professor Crocker's work over the last five years, but not a single citation to either of the two studies containing this estimate. Nor as far as I can determine have there been more recent estimates developed, based on Crocker's estimate. Perhaps this condition of nonoccurrence of a dialogue is due to some lack of credibility for such an estimate. But also it may result from the lack of development of public, widespread indications of substantial benefits from acid reduction, of the type of indications that one had available in automotive accident rates, smog emergencies and asbestos related disease rates.

Even more basic, the $5-billion estimate is not reproducible. As shown in Professor Crocker's Table 1, it is a sum of five separate estimates. The smallest of the separate estimates can be reduced to zero for reasons that Professor Crocker now provides. But by giving away the $250-million dollar benefit, as an inaccurate estimate, it is not possible to keep the other, separate $2-billion estimates. The largest category of benefits, from reducing materials corrosion associated with acidification, cannot itself meet the most elementary test of replicability. Another analyst besides Professor Crocker cannot take the population in the eastern half of the United States, multiply it by the per capita costs in Sweden from materials corrosion, and obtain $2 billion as an estimate of cost to be foregone from reduced acidification. After seeking assistance along these lines, I was offered the suggestion by another analyst that you would have to divide by the ratio of superbowl football contests won by American Football League representatives to the total of all such teams. Without descending to that degree of sarcasm, it is necessary to conclude that the estimate of costs foregone or benefits to be realized is purely a judgmental number derived from the author's own experience and imagination.

Even that estimate would be useful if it were focused directly on a policy issue. But with respect to materials corrosion and forest ecosystems benefits, comprising a substantial proportion of the total, the alternatives relevant for policy would be five, ten, or fifteen million tons of sulfur emissions. Those dealt with in the estimation process are no acidification or present levels of acidification. This is not a reasonable set of policy options, so that the estimate, whether or not judgmental, must be substantially irrelevant.

Then are we left with no estimates of benefits, or an estimate of no benefits, with respect to acid rain policy? The answer must await further research. There has to be additional cross-sectional analysis undertaken of the impact of acidification on regions and on counties within a specific region of the country. Studies of this nature, whether done by epidemiologists or economists, require millions of dollars of outlays on data collection and analysis. My position is that these studies should be undertaken, because their benefits are substantially in excess of their costs, even though I expect that they will show no substantial benefits from reduced acidification.

Indeed, it is because such findings are the predictable results that these studies have not been undertaken by the Environmental Protection Agency and related interest groups at the present time. That result would require the advocate agency to face the fact that no new, large-scale acid rain reduction program can be justified at this time.

DISCUSSION

THOMAS CROCKER: Frankly, I disagree only a little with respect to Paul's comments. The substance of what he presents is mostly correct. As he states, the $5-billion figure rarely appears in the technical scientific literature; it does not appear because it is simplistic. It has, however, frequently been referenced in the mass media and by politicians of various stripes. It has sometimes even been mentioned in the technical environmental economics literature, albeit in a very, very cautious fashion and without any commitment as, for example, by my good friend Rick Freeman. His approach to the figure is entirely appropriate.

Paul MacAvoy appears to miss the main argument of my paper, however. He is saying that he can invent a totally different set of stories, each element of which will generate a number different than $5 billion. With respect to the acid deposition control benefits issue, I will bet that I can invent a good many more stories than he can. Each of these stories will also generate a different number, and then one can continue the game ad infinitum by inventing stories about how each of these new numbers could be wrong. The question I am asking in the paper is whether any truth value can be attached to numbers derived from parameter values for which no analytical or empirical evidence whatsoever exists. My answer to the question posed, and my answer to Paul's refrain, is "no." A multitude of stories about the connections among parameters, a near-infinite variety of parameters from which one can choose, and parameter values going from negative to positive infinity make the individual probabilities vanishingly small of the stories that Paul would apparently like to tell. The $5-billion story that I have told is based only on parameters generally acknowledged to be influential and for which there exists some fairly accurate and precise evidence about magnitude. So it's a question of whether you want to believe a MacAvoy story, or stories, I should say, or if you want to believe the particular story that I told which I don't entirely believe myself. (Always hedge a little.)

It is nevertheless true that my story could be improved upon with only modicum of research investment. With a $10 million or even less, Rick Freeman, for example, could undertake a 5-year project that would show very precisely how the economic consequences of acid deposition vary with acid deposition levels. He, or his equal, would not necessarily even have to wait around for the biologist and other natural scientists to provide him dose-response functions. For example, some of his research results would undoubtedly refer to the behavior of producer cost functions which he had derived using modern duality techniques. Given a few conditions, the production functions or dose-response functions which must underlie these cost functions can be inferred. Thus, not only might his economic results be used to a degree to direct natural science research efforts, they might also obviate the need for some of these efforts.

The aforementioned research results would probably also provide many opportunities for biologists, economists, and other interested parties to invent yet more stories, which would be said by some to require yet more research before any decision could be made. The obvious basic question is how much precision and accuracy in estimates is enough. I argue that, given the flabby form of the current available information on economic consequences, very substantial gains in precision and accuracy can be acquired by some modest but judiciously selected research investments. Materials impacts upon buildings and other urban structures are my favorite candidate.

Dean MacAvoy concludes that a free market in research ideas and results will satisfactorily resolve these questions. Because most North American acid deposition research is funded by program agencies (e.g., USEPA and Environment Canada), I have grave doubts about the validity of his free market premise. It is highly questionable whether these agencies are willing to fund the long-term research and to lay out all the analytical complications associated with the stories he tells or would like to have told. In these agencies, there is a strong propensity to engage in fire-fighting. I am unaware of any alternative hypotheses that can be used to explain the fascination USEPA has had with aquatic ecosystem impacts and its relative neglect of materials impacts. Paul implies that those who allocate research funds have no influence over the research programs adopted, and that the paths taken in subsequent research programs are independent of the earlier programs. Even people who cannot swallow a formal Bayesian view of the world will find this position to be somewhat whimsical.

Nevertheless, with respect to Paul's general theme and his desire for analytically and empirically complete economics research that is well grounded in the natural sciences, I could not agree more. However, for the reasons that I have just stated, I question whether that funding in any area of environmental economics is likely to be forthcoming. More importantly, I seriously doubt that it is worthwhile always to fulfill Dean MacAvoy's, and my, desire.

JACKIE TUXILL (New Hampshire Citizen's Task Force on Acid Rain): It isn't so much a question as a comment that I feel I have to make. And this is directed at Dr. MacAvoy, and I have to say that I strongly object to the put-down to the seriousness of the problem that we are here to discuss today. The seriousness has been attested to by numerous distinguished and scientific bodies and I'm not going to enumerate them right now. One reason that detailed economic information is not presently available is because of the complexity of the problem that we are addressing. You talked about mesathelioma that is caused by asbestos. We are talking about numerous problems with respect to our society today, about ecological effects on forests, and on aquatic systems; we are talking about materials damage; we are talking about damage to water supply; and I just have to say that I object to the type of put-down that you made to the problem.

I cannot begin to argue with you economically, because I am not an economist. But I do have background in natural resource management; I have background in ecology and I will say to you that when we are being faced with a potential threat to an ecosystem such as the forest ecosystem, when we are being faced with potential threat to our water supplies, I can't say that I can provide you with the numbers that you say you feel you need, but I assure you that this is a serious problem and it cannot be put down the way you are doing.

EFFICIENCY AND EQUITY: CAN ACID RAIN POLICY INCORPORATE BOTH?

EFFICIENCY AND ENVIRONMENTAL POLICYMAKING

Paul R. Portney

Senior Fellow
Assistant Director, Quality of the Environment Division
Resources for the Future
1616 P Street, N.W.
Washingotn, DC 20026

Any time one has an assignment to explain what economic efficiency is, it is always best to start by saying what it is not. I'd like to do that here.

The first thing to note is that pursuit of economic efficiency is not an attempt to ignore things that don't get bought and sold in observable, organized markets. There are some economists who are guilty of creating the impression that it is; perhaps there are even some economists who really believe that that is what economic efficiency is. But to the trained professional benefit-cost analyst, that's not what economic efficiency is.

Nor is it an attempt to shoe-horn all of the positive and negative effects that attend regulatory or other kinds of government policies into a dollar and cents calculus. That is to say, even if one takes into account some things that don't get traded in markets, economic efficiency does not necessarily require, although it often implies, that it's handy to value these things in dollars. It is possible to make some kind of rough or qualitative benefit-cost trade-off (that is, to try to determine economic efficiency), with some things on each side of the ledger that are monetized in dollar terms and some things that we can't translate into dollar terms.

Finally, economic efficiency is not necessarily an attempt to eliminate matters spiritual or ethical from decisionmaking about environmental regulation or other kinds of things. It becomes more difficult to take values like this into account in doing benefit-cost analyses and, in fact, it may be impossible to quantify such things in benefit-cost analyses. But economic efficiency and the pursuit of benefits in excess of costs in principle allows us to list things qualitatively and attempt to give them at least qualitative weight in deciding whether or not we will want to pursue a certain policy. Now, having given you my view of three things that economic efficiency is not, let me try to give you some idea of what I think it is.

As many of you know, a formal definition of economic efficiency, or of an economically efficient state or allocation of resources, is one in which no reallocation of resources is possible that would make one person better off without making at least one other individual in the society worse off. In other words, there is no way to reallocate financial, environmental and other kinds of resources in such a way that there is not at least somebody who is made worse off as a result of that reallocation. Speaking formally, that condition or state is also called Pareto optimality. Now informally, and for the non-economist, an economically efficient policy change can be said to be one that causes more good than harm.

On these grounds, I am saying that benefit-cost analysis or the pursuit of economic efficiency in government policymaking is really no more than common sense. When we try to make a decision about whether or not to move from Washington to Cincinnati or buy a car, or build a new wing on our house, we try to list all of the advantages and disadvantages associated with each option. If some of them can be expressed in dollars and cents, we can net those out easily and we try to decide whether or not the good effects outweigh the bad effects. Believe it or not, that is more or less what economists are up to in the pursuit of economic efficiency in regulatory and other kinds of policymaking.

A final point I want to emphasize is that, contrary to the belief that efficiency suppresses or crowds out ethical or spiritual values in decisionmaking, by making it possible for us to reduce the costs of meeting certain goals, it can enable society to have more resources to pursue these other, often nonmaterial goals, for example, to take the gains in the form of enhanced leisure which enables us to sit by the side of a lake and contemplate.

I have presented the position that economic efficiency is really no more than the formal pursuit of common sense in government decisionmaking. If that is the case, why is benefit-cost analysis and the pursuit of economic efficiency so controversial? Let me tick off a couple of reasons why I believe this is so, all of which bear on economic efficiency as applied to acid rain policy.

Benefit-cost analysis and the pursuit of efficiency are controversial because they use individual's preferences as a means of valuing benefits or costs. One has to wonder whether, in a society where Brent Mussberger is paid $2 million a year to announce sports for CBS, but which would reward Mother Teresa at a very low rate if she were to work in the market, we want to use market prices and individual preferences as a basis for guiding decisions in environmental and other kinds of policy. The problem with any other approach is that if we don't let individuals value effects as they feel them, whose values will we use? If we were to use mine, I might regard that as an improvement over market valuation. If you were to use somebody else's who felt that what I do on a daily basis was worth 35 cents a year, I would take exception. Thus, individuals' valuations are used because it becomes very difficult to know whose valuations to use otherwise to assess benefits and costs. Nevertheless, that's one of the problems associated with the pursuit of economic efficiency.

Another difficulty that arises regards characterizing a program as economically efficient if the gains to the gainers exceed the losses to the losers. If the gainers do not compensate the losers, some individuals will be worse off even under an efficient policy. Those individuals can be expected to complain loudly. If, as is often the case, costs are concentrated on a very few individuals, while benefits are diffused widely over the population, the losers may be able to scream loud enough and use the political process to stand in the way of efficient policies. I'll say a word below about how inefficient policies often manage to make it through our governmental system.

Another problem with economic efficiency and its pursuit is, "What gets counted?" This goes back to my earlier point that, in principle, all positive and negative effects should be counted, whether they get traded in markets or not. It's sometimes the case that analysts suppress the non-marketed things because they have difficulty valuing them. Indeed that's why these things don't trade in markets sometimes, because there is no way to appropriate those benefits in such a way that, by providing a public good like environmental quality, one can get rich doing it.

A fourth problem in the pursuit of economic efficiency is the comparison of the positive and negative effects. Basically, this is the apples and oranges problem. It leads to the belief on the part of some that economists try to shoehorn everything into dollars. Benefit-cost analysis does not require that. All that is required is some kind of common denominator to weigh the plusses and minuses. This common denominator could be pears, roses, salt, or just about anything so long as we have the ability to translate positive and negative effects into this common denominator. Why is it that we end up using dollars most of the time? Well, because that is a very common way to express the costs of a policy. Since the "con" side is very traditionally expressed in dollars, it's become fairly common to try to express the benefit side in dollars to the extent possible, but it is no doubt one of the difficulties that arises in benefit-cost analysis.

A final difficulty with the pursuit of economic efficiency and the use of benefit-cost analysis in evaluating regulations has to do with the problem of discounting benefits and costs, which often inconveniently accrue at different points in time. So long as that's the case, we need to find a way to reduce both streams of benefits and costs to present values. Unfortunately, there is much disagreement about which rate to use to reduce future benefits and costs to present values. The most common approach is to use a variety of discount rates to see how sensitive the calculation of benefits, costs and net benefits are to the use of a particular discount rate.

Permit me one last introductory observation for those who feel that there is something inherent in benefit-cost analysis biasing it against environmental or other regulatory programs. Reviewing a number of recent decisions by federal regulatory agencies, including the lead phase-down at the Environmental Protection Agency, the particulate standard likely to emanate from EPA soon, the National Highway Transportation Safety Administration's decision to require a third taillight mounted high on the trunk of a car, and NHTSA's air-bag decision, I find that in each of these cases, the pro-regulatory option was driven by benefit-cost analysis. While one can certainly find cases where benefit-cost analysis has tipped decisions against environmental programs, there are also any number of examples where sound environmental programs are supported by benefit-cost analysis.

Let me now turn to efficiency and acid rain. I'll begin with a ridiculously easy example. Let me hypothesize that annual SO_2 emissions in the U.S. could be reduced by about 50 percent, or about 12 million tons a year, for the sum of $50 million. Comments by the earlier panelists notwithstanding, I would say that even if we couldn't translate acid rain control benefits into dollars and cents, most people would have the feeling that the likely benefits to aquatic ecosystems, exposed materials, forests, soils, and perhaps even human health are likely to be in excess of $50 million a year. In that case, even though we might not be able to translate benefits into dollar terms, I believe most people would be willing to run the $50 million a year gamble and go ahead with a 50 percent reduction for $50 million. If so, that decision would reflect a belief that expected benefits are likely to exceed $50 million.

Ah, but the real world raises its ugly head. What are the actual costs and benefits of the real acid rain control policies that have been before Congress? Relying on ICF's cost estimates, if a 10 to 12 million ton per year reduction in SO_2 emissions is to be considered, the costs are likely to be $3 billion annually, perhaps as high as $6 billion depending on whose estimates are used. Recent estimates seem to hover in the neighborhood of $3 billion for a 10 to 12 million ton a year reduction. A $3-billion increase in costs to electric utilities would mean rate increases in some parts of the country on monthly electricity bills of 10 to 20 percent. This is unless there were some kind of cost-sharing that ended up subsidizing rates in those parts of the country that would be particularly hard hit.

What about the benefit side? It would have to be quite substantial if we are thinking of spending between $3 and $6 billion a year on control costs. Let me talk about some of the uncertainties as I see them. One of the problems is that if SO_2 emissions are reduced by 10 to 12 million tons a year, there still exists uncertainty about how much reduction in sulfate deposition will result. More importantly, there is even greater uncertainty about where sulfate deposition will be reduced. In other words, if emissions are reduced in the Ohio River Valley, where will acidic depositions be reduced? On a question like this, which is obviously much more scientific than economic, there are valuable experiments being conducted that will provide important information. Examples are CAPTEX, also known as the cross-Appalachian tracer experiment, and the MATEX, or massive air tracer experiment, soon to be conducted by the Electric Power Research Institute and a number of governmental and nongovernmental organizations.

Second, even if we know where the deposition will fall, in some cases it is difficult to know the effect on the pH of the receiving soils and waters. This depends at least in part on natural buffering capacity, and is another uncertain element, although research is providing us with more and more information all the time. Finally, even if we knew what the change in pH would be, uncertainty would remain concerning the physical effects on ecosystems, soils, forests, materials, and crops, given a change in the pH of the rain that falls on them or in the water bodies nearby.

Finally, suppose we knew all of the beneficial physical effects of reduced acid deposition. What could we say about valuing these benefits? First, while a lot of benefits from acid rain control would be difficult to value, at least some of the benefits would not be. Thinking about reductions in agricultural output, these are products that trade in organized markets. If we knew, and this is a big "if", that control efforts would result in so many more bushels of soybeans, or corn, or wheat, or whatever, then we could use market prices to value that additional output. The same thing is true with some materials damage. If one must replace awnings less frequently, or repaint less frequently, these, too, are things that trade in markets and reasonably good information is available on how to value these changes.

Obviously when one gets into ecosystem damage, possible human health damage, or damage to areas which may be useful for recreation, valuation becomes more problematical. We have a long way to go. As a result of the uncertainties I have identified here, it is my opinion that it's unlikely over the short term that we will have reliable estimates of the benefits associated with the acid rain control programs that are before Congress now. That's not to say that we might not be able to make ball-park estimates. But I think we have a long way to go before we will have benefit numbers that are sufficiently reliable so that we wouldn't feel guilty about comparing them to the cost estimates.

One might say, you've talked a lot about economic efficiency, and costs and benefits, and now you have alleged that it's going to be very difficult to come up with benefits numbers. Does that mean that economic analysis or policy analysis can really tell us nothing about the decisions that are before us on acid rain? My answer to that is "no". I say "no" because we can call in the "country cousin" of benefit-cost analysis: cost-effectiveness analysis. By cost-effectiveness analysis, I mean comparing the costs of meeting a predetermined goal and selecting the least costly way of reaching that goal. What I am saying is that if, for political or religious or scientific or other reasons, Congress decides to adopt an acid rain control program, we can invoke the old axiom that there is more than one way to skin a cat, and turn attention to the job of minimizing the costs of meeting a predetermined SO_2 reduction program. That's what is meant by cost-effectiveness analysis. I am going to confine the rest of my remarks to the use of cost-effectiveness analysis in the context of some of the acid rain control programs before Congress.

It is at this point that I begin to antagonize potential inventors because the next thing I want to say is that, given the magnitude of the acid rain control programs that have been before Congress over the past couple of years, there are basically only two ways in the short term to get the SO_2 emissions reductions that those bills have called for. One is through forced scrubbing or through a technology-based approach; the other is a fairly substantial shift into lower sulfur coals. I say this antagonizes inventors because I am sometimes sent blueprints with a letter from someone with a device which is alleged to be capable of reducing 12 million tons of SO_2 for $7.55. These claims can be immediately discounted although there are developing technologies to reduce SO_2 emissions. These include both advanced fluidized bed combustion and limestone-injection multi-stage burning (LIMB). In addition, coal washing is a very cost-effective way to remove some sulfur from coal. Finally, the whole debate about acid rain has probably given short shrift to conservation as a potential SO_2 reduction strategy; I hope in future debates about acid rain that more attention is paid to how SO_2 emissions might be reduced through energy conservation as opposed to fuel switching or technological means.

Nevertheless, for the massive kinds of reductions contemplated in legislation, one is left with fuel switching or scrubbing to reduce SO_2. Having said that, it must be noted that the cost differences between a fuel switching approach and a forced technological approach are very substantial, according to best estimates. By a fuel switching approach, I don't mean forced fuel switching. Rather I mean giving the affected power plants and perhaps even industrial sources, the flexibility to meet their apportioned share of the total SO_2 reduction however they see fit. If they choose to close the plant down, they can do that; if they want to reduce sulfur emissions using coal washing or advanced fluidized bed combustion, they can. Basically, however, the source is given a choice among the various emission-reducing approaches.

For a ten million ton a year reduction, or something in that neighborhood, the cost difference between the so-called freedom-of-choice approach and the forced scrubbing approach is on the order of a billion dollars per year, perhaps more. If the real goal of acid rain legislation is to improve the quality of the environment, why not reduce SO_2 emissions using the least expensive means possible? That would seem to be a goal with which we could all agree.

Unfortunately, if the fuel switching approach is chosen, political problems arise. They are the sorts of distributional problems alluded to earlier. Remember that even an efficient policy can leave some people worse off, even though the gains to the beneficiaries may outweigh those losses. In particular, those who are made worse off if a fuel switching approach is adopted are those people who mine high-sulfur coal, primarily in southern Illinois, Ohio, Indiana and western Kentucky, and who will find the market for that coal disappearing as mid-western utilities began to purchase low-sulfur coal from other areas. As everyone is aware, there is quite a controversy about just how many miners would lose their jobs. Some estimates seem to suggest, and I say this only partially facetiously, that if we allow fuel switching to take place, virtually everybody in the United States will lose their job. I never realized that high sulfur coal was so important to the economy ... more important than the auto industry, to hear at least some proponents of forced scrubbing talk about this issue!

Nevertheless, such exaggerations do not obscure the fact that there would be adverse economic impacts in certain high-sulfur coal areas if acid rain controls brought substantial fuel switching. Having looked at much of the evidence, it appears that the best estimate of the potential direct job losses among high-sulfur miners, is about 20,000 jobs by the year 1995. I want to begin to whittle that number down a little bit. Although there may be 20,000 fewer job slots in high-sulfur coal mining areas than there would be in 1995 if the nation took a forced scrubbing approach, that does not mean that there would be 20,000 people who are currently working in coal mines who would lose their jobs by 1995. This is because a significant fraction of miners currently active, about a fifth, I believe, are aged 50 or above. Thus, by 1995 these workers will be 60 years old or older and, hence, candidates for retirement or early retirement. Therefore, some current miners will not lose their jobs even if the fuel switching approach is permitted.

Second, some of the coal mines likely to be affected are certain to be closed down because they are played out. For that reason, the original estimate of 20,000 dislocated coal miners must be further reduced. Hence, 10,000 may be a better estimate of the number of people who are currently working in high-sulfur coal mines who may lose their jobs by 1995 if electric utilities and other sources are given the freedom to switch fuels under an acid rain control program.

To protect these people, whose livelihood, I believe, should be seriously considered in any kind of acid rain program, we have been given the forced technology approach. The problem it creates is that if the nation spends an extra one or two billion dollars a year to protect 10,000 jobs, then it would be spending somewhere between \$100,000 and \$200,000 per year, *per job saved.* I would allege that this is an economically inefficient way to protect the jobs of people making \$25,000 to \$30,000 a year. This does not mean that we should necessarily write these people off. What it does mean is that if we could save a billion dollars a year using a more economical approach to SO_2 reduction, then part of that savings ought to be directed toward creating some kind of compensation program for those adversely affected. This might be accomplished through a direct buy-out program, or a job retraining assistance or relocation program, or any program to lessen the burden on those who would be harmed because the nation is pursuing a least-cost approach to acid rain control. The relatively little writing I have done on the acid rain issue has been addressed to how we might "have our cake", which is to say a lower cost approach to acid rain control, and eat it at the same time in the form of providing assistance to people who would be harmed by a fuel switching approach. That's the essence of cost-effectiveness analysis as applied to this one important aspect of the acid rain debate.

One final word about how much of a role efficiency actually plays in environmental and other kinds of decisionmaking in Washington. The view is rife, especially in the environmental and public interest community, that especially over the past four years, efficiency has somehow reigned supreme. I would allege that while perhaps greater attention has been paid to economic efficiency recently, it is honored much more in the breach than in the observance. Indeed, for the coterie of policy analysts in Washington a favorite watchword is, "There is no constituency for economic efficiency". There exists a band of economists, sometimes woefully misbegotten, I suppose, who try to protect efficiency concerns in environmental decisionmaking. But it is not an easy battle and, 99 times out of a 100, efficiency will give way to political concerns. The latter often create a situation such that, if costs are concentrated on a small number, they will outweigh diffuse benefits even if the total of benefits is far in excess of costs.

In spite of this somewhat discouraging assessment, proponents of economic efficiency in environmental decisionmaking should not despair. In both the Carter as well as the Reagan EPA (at least in the Ruckelshaus regime), the sophistication of the analysis accompanying proposed and final regulation has been high and is improving. Moreover, EPA has finally taken the lead in promoting the use of economic incentives in regulation. I refer here to the "bubble" and "offset" policies, both of which could find valuable application in any solution to the acid rain problem. Thus, there is a record of at least some concern with efficiency in rulemaking; it can and should play a role in the acid rain debate, but will not be the only or even a major consideration in the final shape of any proposal that may emerge.

DISCUSSION

RUTH GONZE (American Public Power Association): Would you comment on the view that if revenue-raising or cost-sharing mechanisms are adopted to spread the cost of acid rain controls, emissions taxes are the way to do it.

PAUL PORTNEY: Yes, I have said that we ought to give some thought to effluent charges as a way of raising revenue if we decide to create a kitty to compensate those parts of the country that would be especially adversely affected under an acid rain control plan. I would be more enthusiastic about their use but for one disadvantage in this particular context. If they are used to create a kitty to subsidize either the higher costs of low-sulfur coal or the installation of forced technology, the people paying into the fund will be those expecting to be subsidized by it. Thus, we would basically be taking money out of their right pocket and putting it into their left pocket. Therefore, the use of effluent charges alone is not as attractive as would be a straight effluent charge without mandated technologies or a need to raise revenue.

TOM ULASEWICZ (Adirondack Park Agency): I'm a little bit concerned that the rather overwhelming emphasis in this discussion is on SO_2 reduction as a response to the acid rain problem. We heard this morning that nitrogen oxides play a role in certain circumstances. I think I'd like to make a plea that nitrogen oxides be taken more into the picture and maybe we could talk not so much about ten or twelve million ton SO_2 reductions.

PAUL PORTNEY: I think that your point is very well taken. Let me say why I think it is we have tended to concentrate on SO_2. Three reasons come to mind. First, at least to date, we have conceived of the acid rain problem primarily as a problem in the 31 eastern states, although there obviously is some evidence to suggest that certain parts of the West have been experiencing some serious acid rain or acid fogs. Because we have concentrated on the eastern 31 states, we would naturally concentrate attention on SO_2 because it is thought that it accounts for about two-thirds of the acid deposition problem in the East. That's one reason. The second reason is that it's my understanding that even at large stationary sources, when compared to SO_2 removal we know less about technology for NO_x removal. The third reason is that if you begin to concentrate too much on NO_x, then you bring cars into the acid rain control issue. Since there are 140 million cars on the road, you begin very quickly to multiply the number of sources that are involved in some kind of control program. That's not to say that there might not be opportunities to reduce the cost per unit of acidity removed, but there are also a hell of a lot of vehicles to be controlled which creates political problems that a lot of people here understand much better than I. I think those are the 3 reasons why we tended to focus on SO_2, but I think the point is well taken that we ought to at least consider the possibility that in reducing total acidic deposition perhaps there are some attractive NO_x control opportunities and those certainly should be factored into any kind of thinking about a national acid rain control program.

DESIGNING EFFICIENT, EQUITABLE POLICIES TO ABATE ACID RAIN

Lester B. Lave

Professor of Economics
Carnegie-Mellon University
Pittsburgh, PA 15213

INTRODUCTION

Economists recognize a number of criteria for good policy: economic efficiency, equity, administrative simplicity, transparency of the objectives and methods of the program, etc. (Lave, 1981). Thoughtful economists will admit that each of these criteria are generally important in deciding which policy to choose. However, most economists will either confess or assert that economic efficiency is the first and most important criterion, at least in the sense that a program should be economically efficient, whatever else its attributes.

The reason for the focus on efficiency is simple: Who wants to waste resources? If an objective can be accomplished for lower cost, why pay more? Even if one has no good idea of what one wants to do, it makes sense to do whatever is to be done efficiently. What it comes down to is an almost Calvinistic devotion to preventing waste.

All of the other criteria are concerned with gaining acceptance of a program, being able to implement it well, or selling it to society in general and politicians in particular. While certainly relevant, these other criteria lack purity and theoretical clarity. They are each subject to differing interpretations and even arguments about what actions are better than others.

This devotion to efficiency reached its height in the early part of the last century when economists (under the guise of utilitarianism) believed that efficiency could be used to determine good policy, without any need for qualification by the other criteria. Thus, the benefits and costs of each policy could be evaluated and one could find a unique best policy offering the greatest net benefits to society; whatever criteria were relevant were already incorporated into this policy and there was no need to compromise efficiency.

This devotion to efficiency came under attack in neoclassical economic theory and political economy (Mishan, 1965). So far as economists were concerned, these notions were discredited as being both incomplete and as needing assumptions unlikely to approximate conditions in the real world.

But these ideas didn't die. Like the Phoenix, these ideas have sprung from the ashes and been reborn in Executive Order 12291. President Reagan directs each agency to choose the policy with the greatest net benefit or give a written explanation why this policy was not chosen. It is a mistake to take the intellectual foundation of the Executive Order too seriously; rather it should be viewed as a convenient device for asserting presidential authority, via the Office of Management and Budget, over the decisions of the regulatory agencies. According to this directive, agencies must examine all proposals regarding acid deposition, quantify the effects, translate each of the benefits and costs into dollars, discount them to the present, and choose the proposal with the greatest net benefit. Indeed, the Executive Order requires EPA to follow this procedure unless it is precluded from doing so by legislation.

EFFICIENT IMPLEMENTATION

A program that is efficient in this sense, should be implemented efficiently. Indeed, the assumption in this type of cost-benefit analysis must be that each of the alternative policies evaluated are implemented efficiently. After all, why should you spend more implementing a program than is necessary? There is an extensive economic literature on how to implement a program efficiently (Baumol and Oates, 1975, Kneese and Schultze, 1975). Basically, efficiency requires that each person should be given as much leeway as possible to adapt the policy to his specialized circumstances.

Indeed, even if the overall policy is not chosen with efficiency in mind, it still can be implemented efficiently. Whatever your goal, why not implement it at least cost? Thus, one has the anomaly of asking for efficient implementation of a program that is not seeking to serve an efficiency objective.

CAN BENEFIT-COST ANALYSIS ENCOMPASS ALL ISSUES?

Before proceeding to consider the other criteria and to apply those notions to acid rain, I want to explore these ideas a bit more.

Using a benefit-cost analysis to choose a policy assumes that all the important aspects of a policy can be quantified and valued in the benefit-cost analysis. For example, damage to commercial forests is easily valued and not conceptually difficult to quantify. Changes in the ecology of unmanaged forests will be much more difficult to quantify and may be impossible to value (Freeman, 1979, NSF, 1984). Yet, if changes in forest ecology are potentially important in choosing a policy, both the quantification and valuation must be done with confidence if the benefit-cost analysis is to arrive at the best policy.

There are economists who go further and assert that the additional criteria of equity, administrative simplicity, transparency, etc. are not independent criteria, but actually can and should be handled within the benefit-cost analysis. For example, if the "wrong" people are paying for pollution abatement, one need merely quantify and value the disbenefit of having a particular group pay and add this to the cost side of the particular policy alternative. Thus, a policy that appeared to have the greatest net benefits without accounting for equity might not look as good once equity is quantified and valued. Fine, this is simply generalizing benefit-cost analysis to account for all the relevant effects.

If one could quantify the other criteria with confidence and could figure out how to value the effects, there would be no conceptual difficulty associated with putting these effects into the benefit-cost analysis. However, if this cannot be done with confidence, the benefit-cost analysis can be misleading and analysts will be pretending to do something they cannot.

HOW IMPORTANT IS EFFICIENCY?

For a complex issue, such as abating acid rain, I don't think any economist would assert that benefit-cost analysis can incorporate (with confidence) estimates of all the important issues. Personally, I would go further and assert that it is conceptually impossible that benefit-cost analysis ever could handle such complicated issues. But for now, it is enough to assert that EO 12291 is pernicious here since selecting the alternative with the greatest net benefits would only by chance select the alternative that is best from the standpoints of equity, administrative simplicity, and transparency. Indeed, I would be skeptical even that the alternative which the policy analysts asserted had the greatest net benefit was truly the one with the greatest efficiency, because of uncertainty.

Policy alternatives should be evaluated with efficiency, that is, benefit-cost analysis, as one input; but efficiency should not be the only or even the dominant input. I want to be clear that I think that efficiency is a very important criterion, but it is not the only criterion. Insofar as some aspects of efficiency can be quantified with confidence, this information is valuable in making a decision. However, if the benefit-cost analysis is little more than guessing, or if it manages to address only a tiny portion of the issues, it may simply mislead. Nonetheless, while decision makers must be educated about how to use and interpret benefit-cost analyses, I believe that use of this tool will lead to better decisions.

How important is it that a policy be implemented efficiently? Much of my skepticism disappears on the implementation question. Although there are the same sorts of objections, I feel much more comfortable asserting that implementation can and should be made more efficient. Certainly there have been occasions when economists have unwittingly emasculated a program by preaching efficiency in implementation. And certainly there are important values that get lost in maximizing measured efficiency in implementation (or minimizing the measured costs of implementation). However, these losses seem generally small beside the huge savings in implementation costs and the ancillary and more general gain of keeping the program on target.

EQUITY

If equity were well defined and easy to measure, one could probably include it satisfactorily in a benefit-cost analysis. However, there is not general agreement as to who should pay and who should benefit. Equity is more a family of attributes than a single one. Who benefits and who pays each have myriad facets, including timing (short, intermediate, or long term), geography (generators versus receptors), demographic group (concern for damages and jobs depend on age, sex, race, income, and education), the preferences of those people yet unborn, and valuing jobs and regional development relative to environmental effects (Gianessi et al., 1977). There are also interactions with other concerns, as with other environmental insults such as air and water pollution or endangered species.

In earlier times when there was more general agreement on social goals, due largely to greater cultural homogeneity, there was more agreement on how to define equity. Similarly, within a philosophical or religious system, equity about most matters is well defined. As a polyglot nation with many ethnic, religious, and other groups, there is little agreement on what constitutes equity beyond some basic principles of western culture.

One could define equity via political power, the ability to enact a program that gives benefits to one group or makes another pay. While an examination of recent legislation gives a basis for predicting what Congress will consider equitable, the idiosyncratic nature of these compromises means that this approach does not provide a stable basis for predicting what will be considered equitable in the next legislation.

Thus, we are left in an unsatisfactory position. Equity is considered generally to be the most important criteria, but there is no general agreement on what constitutes equity. In practice, this means that all parties will appeal to equity in terms of some principles that sound at least remotely plausible. In practice, this disagreement will tend to nullify the importance of this criteria.

ADMINISTRATIVE SIMPLICITY

Administrative simplicity includes both minimizing the overhead associated with a program--the administration cost both to the public and private sectors--and keeping the program simple, without myriad other purposes. Any program to be administered across several states, affecting many companies and individuals, and administrators at three or more levels of government must be kept simple, if it has any hope of working. Thus, this criterion refers not only to saving resources but to designing a program that could be successful. Indeed, one way of emasculating a program that cannot be defeated legislatively is to make it so complicated that it cannot be implemented.

TRANSPARENCY

This criterion is in one sense a special case of the previous one. The program should be simple enough so that its purpose and methods are transparent to the general public, to enforcement officers, corporate officials, and to federal judges. Implementation and enforcement are enormously simplified for a transparent program.

However, this criterion is also important in a political sense. The general public and politicians must be able to understand and scrutinize a program. Otherwise there is no real way of checking it. This transparency helps the legislative process and helps prevent future amendments which are attempts to weaken the program.

SOME COMMON PROBLEMS IN ACID RAIN PROPOSALS

The lack of attention to efficiency, and to analysis more generally, leads to some ubiquitous problems.

Focusing on the Biggest Contributor

The first stems from the desire to be where the action is. Thus, if power plants in the Ohio Valley contribute the most pollution, they should be the focus of abatement efforts. While this is a reasonable first approach, it is no more than that. If the cheapest and most equitable way of abating pollution is going after the biggest sources first, that is fine. But often, that is not true. For example, in the case of the Clean Air Act, Congress decided to focus on new sources rather than on the major current polluters.

Wait Until the Scientific Facts are Clear

There is a great deal of uncertainty about the precise source of the acid rain falling on a particular watershed and about how much abatement of which pollutant where would mitigate various observed effects. Since it is also true that controlling sulfur dioxide emissions is expensive, there are many proposals to put off action for perhaps a decade until the scientific facts are clear. Unfortunately, putting off action is a decision to continue present levels of damage. While it seems certain that a sounder, more cost-effective control program could be designed in a decade, this would mean an additional 5-10 years of environmental damage. The decision to delay action is an important decision that requires scientific data just as much as a decision to begin a program now. Since scientific certainty is not available, society has no option but to make the best decision now, in view of current knowledge and conjectures.

Get 'Em While You Can

There is a general tendency to mandate really stringent emissions control for any power plant or category of polluters that get regulatory attention. While attention is limited and there is inadequate time to focus on all polluters, this approach is generally pernicious. By placing what is a punitive burden on those who receive attention, regulators motivate polluters to do anything to escape attention. In any case, there is no reason for polluters to cooperate or be reasonable. They might as well use legal and political challenges to escape attention or at the very least delay the time when they must take action.

Old Source - New Source Bias

Since it generally costs more to retrofit a source to achieve a stringent standard than to design a new source to meet the standard, there are good reasons to impose somewhat more stringent standards on new sources. However, if this bias becomes extreme, it inhibits the building of new facilities and induces owners to keep old facilities which are highly polluting in operation much longer than had been planned (Gruenspecht, 1982).

Cost Is Not Relevant

In its extreme form, this slogan states that environmental goals are to be pursued no matter what the cost. Even short of this extreme, if environmental programs are enacted and implemented without any concern for cost, there will be absurd results that can be used by clever opponents to embarrass the program. Since the public has not thought deeply about these issues, it is not difficult to get contradictory opinions. In one survey most people will say that the environment should be protected without relevance to cost, but another survey could find that most people think that spending $17,000 per fish is absurd. Most people do believe that some cost levels are absurd, even though they support a clean environment. Even if one doesn't accept a benefit-cost calculus for the minutia of decisions, there is still clearly a sort of global rule of reason that requires a rough relationship between benefits and costs for an acceptable policy.

REVIEW OF ACID RAIN ABATEMENT PROPOSALS

These preliminaries out of the way, it is safe to remark that few of the legislative proposals for curbing acid rain have accorded economic efficiency or administrative simplicity much weight. The legislative proposals have focused on two issues: selecting actions which seem necessary to mitigate or eliminate damages, and identifying who should pay the cost. Unfortunately, there is little agreement on either aspect and thus a plethora of proposals.

Knowledge is still rudimentary concerning how great the acid rain problem is and how much abatement, where, would eliminate various aspects. There is disagreement on whether only sulfur oxides need to be abated or whether nitrogen oxides and photochemical oxidants also need to be curtailed. There is disagreement on what the effects would be of abatement of each pollutant individually or all together, as well as about what locations need to be the focus of abatement efforts.

Since there is so much room for reasonable disagreement, there are many seemingly reasonable alternative policies. The error here is attempting to formulate detailed policies at a time when there is so much uncertainty. To be sure, precisely this approach was used in the Clean Air Act Amendments of 1970, especially in specifying automobile emissions standards (see Lave and Omenn, 1981). I don't want to rekindle the debate about whether that was a smart policy at the time, but it is clearly not an acceptable policy in 1984 because of the costs and disruptions.

Pointing out our current lack of knowledge is not the same thing as advocating that no actions be taken until all the facts are clear. However, programs that are extremely expensive and whose efficacy rests on somewhat doubtful assumptions should probably be put off.

Legislative proposals can be characterized in terms of those which view acid rain as a national problem and demand national action in abating sulfur emissions, and those which do not. For the proposals which demand national action, it is unclear whether the motivation is to ensure that everyone experiences some increase in electricity rates due to the effort (spread the burden) or whether the problem is perceived to be the more general one of air pollution, with acid rain providing a focus for action. Actually, I find myself inclined toward this "extreme" position, since I see air pollution as generally damaging and sulfur oxides as particularly damaging, especially after they have remained in the atmosphere long enough to be oxidized into sulfates (Chappie and Lave, 1982).

I feel apologetic for favoring this approach since it appears to be expensive relative to programs more targeted on acid rain in the Adirondacks. However, it is the air pollution problem more generally that concerns me, with acid rain being only a small, not terribly large part. In this view, solving the acid rain problem would be a byproduct of getting on with abating sulfur oxides through the nation. This is an approach which could be executed through the existing Clean Air Act.

Virtually all proposals focus on a narrower area, such as the Ohio River Valley or the 31 states adjacent to or east of the Mississippi River where 83 percent of coal is burned. The proposals differ as to whether all utility boilers (coal and oil) would have to reduce sulfur emissions, or whether abatement would be targeted on the largest plants. The focus is on utility boilers since the economic damage is argued to be smallest there. These utilities are regulated to get a specified return on their investment and so none would fail because of the controls. Furthermore, the costs of the controls would be borne through higher electricity prices. Thus, everyone would bear part of the cost.

Spreading the Costs

The fly in the ointment is that higher electricity prices would hurt the poor and would tend to lead to plant closings and discourage plant location in the area. Thus, this marvelous method of spreading the costs of abatement widely has some problems. One solution would be to go after all polluters, thus putting less of a burden on the utilities. However, as learned under the Clean Air Act, this more general approach would put more pressure on old plants, leading to closure; at the same time, the new source performance standards (NSPS) would put a larger burden on new plants that emitted a great deal of sulfur oxides. Thus, the solution to the problem has not been to look at controls on nonutility sources, but rather to find ways to spread the costs over a wider group of electricity consumers. Thus, the Sikorski-Waxman bill proposes a nationwide tax that would subsidize those who buy electricity from plants that have been retrofitted, which particularly include those in the Ohio River Valley. Thus households in California, Alaska, and Hawaii would be asked to subsidize electricity prices in Ohio.

The other aspect of "who pays" that has caused concern is the loss of jobs that would result if utilities switched to low-sulfur coal, since this is generally cheaper than installing scrubbers. To protect the jobs of miners producing this high-sulfur coal, the Sikorski-Waxman bill would insist that the 50 largest polluters install scrubbers and that no additional controls would be required to burn high-sulfur coal.

Efficiency

The first efficiency issue for these various proposals concerns how much abatement of sulfur oxides is needed where. If the only concern is acid rain in the

Adirondacks, only boilers in a limited geographical area would be required to decrease emissions. If the problem is viewed as encompassing virtually all of the eastern USA, a much wider area is affected. Finally, if the problem is viewed as being national or international in scope, or as being an air pollution problem more generally, then boilers throughout the entire lower 48 states would have to abate sulfur dioxide.

Unfortunately, there is not sufficient scientific evidence to be sure which of these views of the problem is correct. There seems to be evidence of acid rain damage in the southeastern USA, and perhaps some in the West. However, a great deal more scientific knowledge will be required before we can be sure of these problems. There is much uncertainty about the effects of sulfates on health. If sulfates were as toxic as some claim, nationwide abatement would be required (Chappie and Lave, 1982; Wilson et al., 1980; U.S.EPA, 1983). If current levels of sulfur oxides have no effect on health as some claim, acid rain is the principal remaining problem (National Academy of Sciences, 1978). It will be some time before there are confident answers to these questions.

Thus, the most important evidence for answering the efficiency question, namely how much abatement is desirable, cannot be answered with confidence at this time. I believe that convincing evidence exists on the health effects of sulfur oxides, but this evidence has not convinced all of the experts. Thus, a general view, as distinct from my view, is that not enough is now known to answer this question scientifically. If so, a somewhat arbitrary judgment is required about the extent of required abatement.

Equity

The principal equity issues have been set out above. It is important to add that these equity considerations become impossibly costly if the adjustments are continued indefinitely. For example, if electricity users in Hawaii are asked to subsidize those in Ohio for a few years, distortions in the economy should be small. If, however, the subsidy is permanent, the higher prices in Hawaii will lead consumers to shift away from electricity and tend to discourage migrants and new plants. At the same time, the lower prices in Ohio will induce additional electricity use and keep some plants there that should have moved.

This notion is illustrated even more clearly by a policy to replace the income lost by miners when high-sulfur mines closed. For example, if miners did not want to move to find new jobs, and if the relatively high wages meant that the area was not attractive to new plants, government could be asked to continue subsidizing these communities forever. Not only would this be a continuing drain on the rest of the country, it would create a dependent community. To avert these problems, any program to pay for lost wages or lower wages should be phased out within a few years of introduction.

Administrative Efficiency

There are many issues surrounding implementation. With regard to administrative simplicity and transparency, programs must be kept simple. For example, distinctions between old and new, large and small, utility and other boilers all add to complexity and should be eschewed, if possible. Perhaps the simplest rule would be a uniform emissions standard, such as no more than 0.6 pounds of sulfur dioxide emitted per million BTU of coal burned. Another simple rule would be to require scrubbers on all boilers or perhaps on all coal-fired boilers above a certain size. This latter rule appears simpler, since it is easy to ascertain whether the scrubber has been installed; however, this rule is just as complicated to enforce if we want the scrubber to operate effectively while the boiler is in operation. Indeed, if emissions into the air is the criterion for success, some sort of emissions total will have to be measured and enforced.

Another aspect of implementation is the ultimate enforceability. For example, experience under the Clean Air Act showed that if one gave polluters several years to come up with a solution, little or nothing was done until the deadline was pressing (Ruff, 1978). Even then, polluters tended to throw themselves on the mercy of the public and courts, asserting that they could not afford the pollution control, so that the choice was to shut them down or put off the pollution control indefinitely. Since no area cares to lose jobs, some companies have managed to gain years of delay.

This experience suggests that stringent timetables with frequent milestones are needed. Deadlines ought to be pressing at all times so that action cannot be put off. Furthermore, insofar as dealing with the tendency of the polluter to throw itself on the public's mercy, it is better to learn this immediately and try to deal with it immediately. At the time legislation is passed, sentiment is strong enough so that many of these appeals would be denied. Once the public's attention has turned elsewhere, polluters are more likely to receive special treatment.

In any case, the willingness of the public and the courts to enforce the law depends on perceiving that the outcome is worth the cost, that the law is clear and its methods acceptable, and that polluters have been given adequate avenues to take action. This is where transparency is especially important.

One other aspect of implementation is how many parties are required to take action and how many plants must be overseen. Automobile emissions policy has been relatively successful because only a handful of manufacturers were required to take action. Furthermore, testing was done in a few facilities. As the focus has shifted to cars in use, with widespread testing of every vehicle, so the problems of enforcement have become many times more complicated (White, 1981).

Given recent experience, I wonder if Sikorski-Waxman did not have these sorts of notions in mind in opting for requiring scrubbers rather than giving utilities freedom of choice. Requiring scrubbers for the largest utility boilers means there are only a few polluters to focus on. Presumably, having required the scrubbers, the utilities will cooperate in actually using them. This proposal avoids all the problems associated with high-sulfur coal miners losing their jobs. This policy appears to do well with respect to the criteria of equity, administrative simplicity, and transparency. Unfortunately, it is a disaster on efficiency grounds. For a contrary view that once the costs of retraining workers and accounting for costs to the community are considered, a policy that tends to emphasize flue gas desulfurization is actually cheaper, see Jonash, 1984, above.

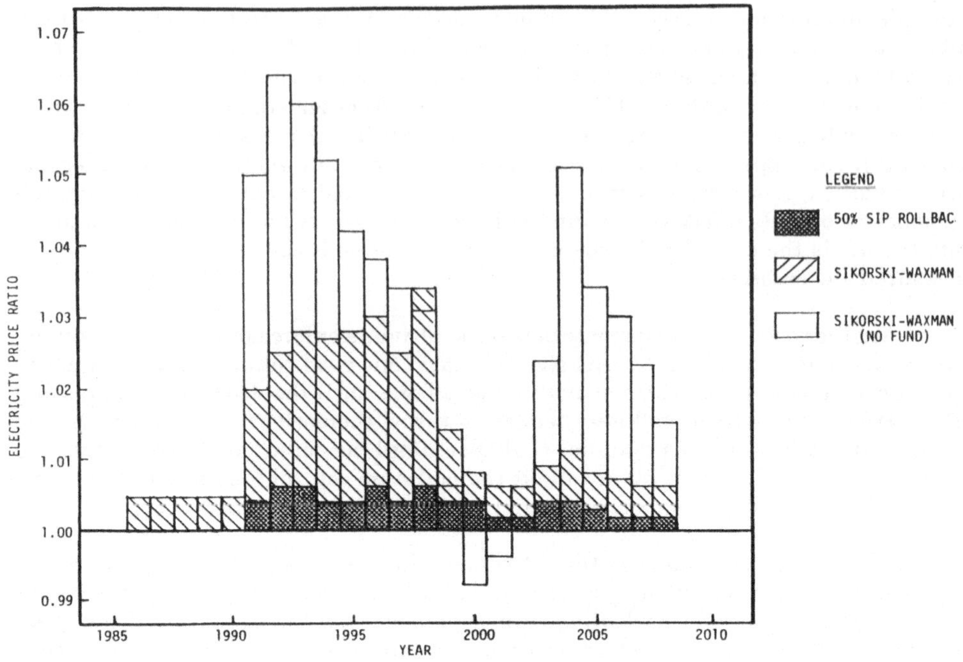

Figure 1. Electricity-Price Ratio Relative to Base Case. Study of Illinois (Bullard and Hottmann).

THE COSTS FOR ILLINOIS: AN EXAMPLE

The magnitude of the cost increases and the costs of the several bills are illustrated in a study of Illinois (Bullard and Hottmann, 1984). Costs are estimated from three cases in addition to a base case of no acid rain legislation. The first case is akin to the Mitchell-Stafford bill and is fitted as a 50-percent rollback in the Illinois state implementation plan (SIP) for the coal-fired utilities. The second is the Sikorski-Waxman bill that requires retrofitting the 50 worst plants (nationally) with flue gas scrubbers and paying for half of the cost with a nationwide tax on electricity. The third is the Sikorski-Waxman proposal, but without any of the costs being borne nationally.

As shown in Figure 1, the results are somewhat surprising. Even in the peak year under the worst case, electricity price is not increased as much as 7 percent. Thus, the cost to Illinois, one of the most affected states, is much less than a 10 percent increase in electricity prices. Even more impressive is the difference between the efficient approach (50 percent SIP Rollback) and the Sikorski-Waxman approach. The efficient approach never has electricity prices rise as much as 1 percent and furthermore the increased prices do not begin until 1990. In contrast the Sikorski-Waxman proposal (without the national fund) raises electricity prices 5 to 7 times as much during the peak years. Even with the national fund paying for half the cost, the Sikorski-Waxman approach raises electricity prices in Illinois by more than three times as much as the efficient proposal. This analysis convinces me that the overall costs of even stringent abatement proposals are not prohibitive and that an efficient proposal is significantly cheaper than a proposal designed to appear more equitable and to be easier to implement.

POLICY DILEMMAS: MOST EFFICIENT IS NEITHER SIMPLEST NOR MOST ENFORCEABLE

The first policy dilemma is that the most efficient (least cost) way of accomplishing the objective is not the simplest, most easily implemented, or most easily enforced approach. Certainly, it is not the one with the greatest equity. For abating 10 million tons of sulfur dioxide per year, the Congressional Office of Technology Assessment (OTA) estimates that the most efficient method would save about 40 percent or $1.5 billion per year over an alternative that seems more equitable and easier to enforce (1984). In this age where no national program or government expenditure that involves less than $10 billion seems worth worrying about, one could be cavalier about this calculation. However, $1.5 billion per year is a great deal of money, resources that could be used for other government programs or used to increase private consumption. For example, it is roughly equivalent to expenditures on food stamps in these states. This is far too many resources to treat casually.

The issue of enforceability is important here. There is sufficient experience with the Clean Air Act to know that programs that are not enforceable accomplish little or nothing. Indeed, they can be pernicious in the sense that one plant acts to curtail emissions and then finds itself at a cost disadvantage relative to rivals that did nothing. This leads to bankruptcy of the virtuous firm, or avoiding that disaster, assures that this firm will never again control emissions until it either sees its rivals actually doing so, as distinct from promising to do so, or is compelled to do so legally.

AN EFFICIENT-ENFORCEABLE APPROACH

If individual utilities and other large boiler operators are to have maximum flexibility, how can we design a program that will be enforceable (Baumol and Oates, 1975; Kneese and Schultze, 1975; Ruff, 1978)? I suspect that a marketable discharge license (MDL) system is the best procedure. Existing boilers would be given MDLs in proportion to current emissions in the affected states. As part of the system, each MDL would have a time limit, and they would expire in a known sequence. There would have to be a large fine associated with emissions beyond the MDLs belonging to the operator. Finally, the burden of proof would be on the operator to demonstrate by a materials balance analysis that its emissions met the MDL limit. The materials balance analysis would proceed on the basis of the sulfur content of the coal burned with deductions for the sulfur removed in washing, flue gas desulfurization, or other removal processes. One might also require some stack tests to ensure that actual emissions are consistent with the calculations.

Care would have to be taken that a firm did not sell all its MDLs and then plead that it would have to fire all its workers unless permitted to continue emissions. That is, there would have to be general agreement among the public, politicians, and judges that this system was fair and that violations actually should be met with the high penalties written into the law.

EQUITY ISSUES

The other issues concern the distribution of who pays for the higher electricity costs and what happens to miners who are laid off from particular mines. There are myriad ways of spreading these costs more widely. However, one question is whether, for example, people in the West will willingly pay part of the increased cost of electricity in Ohio due to the controls. For example, Hawaii has the highest costs of power in the nation. Should they pay still more so that the people in Ohio will not have to face electricity costs that would never be as high as those in Hawaii? Whatever one decides here, it is clear that any redistributive measure should have a relatively short time limit--say a decade. There is no reason or justification for prolonging the transfers beyond such a period.

The same would be said for coal miners. Here the funds for relocation should be of even shorter duration. One would presume either that these workers wanted to stay in their own communities and find some other employment or that they were willing to move and mine coal elsewhere. In the latter case, the costs would be short term and relatively modest. In the former case, the costs are likely to be higher and to persist longer. However, there is also a public policy question as to how much these people ought to be subsidized to stay in their communities, if no industry would relocate there.

WHAT SHOULD WE BE DOING NOW?

Even in the face of major uncertainties there are steps that should be taken now. An examination of abatement costs shows that some steps can be taken at low costs per ton of sulfur dioxide abated, for example, coal preparation to remove pyrite and ash. Indeed, there is a spectrum of abatement technologies and their associated costs. One can identify, at least roughly, the various costs per ton of sulfur dioxide removed and the proportion of the sulfur that is removed from emissions. Despite the uncertainties concerning the precise sources and damages of acid deposition, it is clear that damage to air quality and what is known about acid rain damages are sufficient to justify fairly stringent abatement. Thus, there is virtual consensus that all of the cheap abatement methods ought to be undertaken.

Here agreement ends. Since I believe that health and other effects from sulfur emissions justify stringent abatement even before adding the problems associated with acid rain, I favor stringent abatement without waiting for clarification of the acid rain damages. Those who believe there are little or no additional health benefits to be gotten from further improvements in air quality for sulfur oxides believe that acid rain benefits must be the basis for further abatement. To them, the current uncertainties are critical. These people believe that the more expensive abatement methods will have to await clarification of environmental damages.

CONCLUSION

Although there is vast uncertainty concerning the social benefits of various amounts of emissions control, and some uncertainty concerning the costs of this control, enough is now known to enable both taking action and developing a full program that highlights the critical uncertainties and ways to resolve them.

(a) Coal preparation ought to be required now for all coal being burned in the lower 48 states, whether the coal is burned in an electric utility or commercial boiler, unless the boiler is already fitted with a working abatement system that removes at least 70 percent of the sulfur.

(b) Moderately expensive methods of abating emissions, such as switching to lower sulfur coal should also be undertaken now. Only if the resulting increases in fuel prices are large, more than 10 percent, or if a large part of the burden is concentrated on a particular geographical area, would I be willing to consider taxes or other programs to spread the cost, and even then I would phase such taxes out over a period no longer than ten years.

(c) More research ought to be focused on resolving the uncertainties of health effects of sulfur and nitrogen oxides, since these have much greater implications for social benefits concerning air pollution abatement than current uncertainties about acid rain.

(d) Environmental programs in general and acid rain programs in particular should be justified, roughly, by a balancing of benefits and costs, accounting for all of the current uncertainties.

(e) Implementation of abatement programs ought to focus on efficiency, attaining the objective at least cost, with administrative simplicity, equity, and transparency being important secondary considerations.

REFERENCES

Baumol, W. and Oates, W. 1975, *The Theory of Environmental Policy*, Prentice-Hall, Englewood Cliffs, NJ.

Bullard, C.W. and Hottman, H., 1984, "Strategies for Reducing Acid Emissions from Illinois Electric Utilities: A Preliminary Assessment," University of Illinois at Urbana-Champaign.

Chappie, M. and Lave, L.B., 1982, "The Health Effects of Air Pollution: A Reanalysis," *Journal of Urban Economics*, 12.

Freeman, A.M., III, 1979, *The Benefits of Environmental Improvement: Theory and Practice*, Resources for the Future, Washington, DC.

Gianessi, L.P., Peskin, H.M.and Wolfe, E., 1977 "The Distributional Implications of National Air Pollution Damage Estimates, in: *The Distribution of Economic Well-Being*, F.T. Juster (ed),Ballinger: Cambridge, MA.

Gruenspecht, H., 1982, "Differentiated Regulation: The Case of Auto Emissions Regulation," *American Economic Review*.

Jonash, R.W., 1984, "The Cost of Acid Rain Controls," Arthur D. Little, Inc., Cambridge, MA.

Kneese, A.V. and Schultze, C.L., 1975, *Pollution, Prices, and Public Policy*, Brookings Institution: Washington, DC.

Lave, L.B., 1981, *The Strategy of Social Regulation*, Brookings Institution, Washington, DC.

Lave, L.B. and Omenn, G.S., 1981, *Clearing the Air*, Brookings Institution: Washington, DC.

Mishan, E.J., 1965, "A Survey of Welfare Economics, 1939-59," in: *Surveys of Economic Theory I*, St. Martin's Press: New York.

National Academy of Sciences, Committee on Sulfur Oxides, 1978, "Sulfur Oxides," National Academy Press: Washington, DC.

National Science Foundation, 1984, *The State of the Art of Benefits Assessment*, NSF-PRA.

Office of Technology Assessment, U.S. Congress, *Acid Rain and Transported Air Pollutants: Implications for Public Policy*, 1984, Washington, DC.

Ruff, L., 1978, "Federal Environmental Regulation," in Committee on Governmental Affairs, U.S. Senate, *Study on Federal Regulation*, VI, U.S. Government Printing Office: Washington, DC.

U.S. Environmental Protection Agency, 1983, "Draft Sulfur Oxides Criteria Document."

White, L., 1981, *Reforming Regulation: Processes and Problems*, Prentice-Hall: Englewood Cliffs, NJ.

Wilson, J.Q. (ed.), 1980, *The Politics of Regulation*, Basic Books: New York.

Wilson, R., Colome, S.D., Spengler, J.D., and Wilson, D.G., 1980, *Health Effects of Fossil Fuel Burning*, Ballinger, Cambridge, MA, 1980.

RESPONSE TO LESTER LAVE

Curtis Moore

Associate Counsel
Senate Committee on Environment and Public Works
Room SD-410
Dirksen Building
U.S. Senate
Washington, DC 20510

I agree completely that acid rain is an air pollution problem and an awful lot of people have lost sight of the fact that you can't talk just about acid rain, and that you have to talk about acid rain and acid rain precursors.

It also is not merely an air pollution problem generally; it's also a production efficiency problem. I just returned from Japan and one of the things I saw was a steel mill run by the Kawasaki works where since 1974 they have imposed through retrofit, not through new constructions, roughly a 95-percent reduction in sulfur dioxide, 90-some odd percent reduction of total suspended particulates, and 90-percent reduction of oxides of nitrogen. Those of you who are familiar with pollution control know this is remarkable. In that same period, they have achieved a 93-percent reduction in fuel consumption. I would observe that at the same time we have achieved roughly 95-percent reduction in pollution from automobiles, and we have achieved corresponding increases in fuel efficiency of automobiles so these are not separate problems, and I agree with Lester Lave on that. I do not agree, however, that there is a vast amount of uncertainty in this field. It is true that there is uncertainty, but it is by no means vast. Finally I disagree with the proposition that you should not start with the concept of protecting health. Lester Lave did not say that protecting health is a bankrupt concept, but I think it was inferred from what he said. I don't want to be misunderstood, but if your only consideration is efficiency, it's going to lead not only to some absurd results, but to absolutely inhumane and unjust results. If your only consideration in an area like Medicaid and Medicare, for example, were to be efficient, the most efficient program would be frankly to let the people die. I don't think there is anybody who would suggest that, and I'm not trying to suggest that those people who are opposed to acid rain controls are suggesting that we ought to let the victims of air pollution die, I'm merely trying to illustrate through an extreme example, that we ought first to start with equity and then let efficiency take care of itself.

One of the problems we have here, and this was a point that Lester Lave made, is that all of the analyses are extremely careful to avoid any calculation of the ill-health costs that result from acid rain and its precursors. Yet we know that there is a very substantial body of evidence that those ill-health costs exist. In Japan, right now, in areas where the ambient standards for sulfur dioxide are being met, the

Japanese are paying $370 million a year in compensation costs to the victims of air pollution. However, generally speaking, analysts are loathe to talk about these kinds of health care costs, because they constitute an admission that pollution kills. And yet in fact we know intuitively that it does, and if you want a confirmation of that, one of the interesting experiences the Japanese have had is that in their designated areas of pollution control, where they are measuring ill health because they are compensating people, while they are measuring levels of sulfur dioxide, as the levels of sulfur dioxide have come down, juvenile asthma has disappeared. We know that ill-health costs exist, and we also know that there are some indirect health effects of acid rain.

There are many people who want us to talk about whether we should control acid rain or not and who say that it is a decision which should be made on a cost-benefit analysis. But they want us to use efficiency as they define it. They want us to use efficiency defined in terms of $17,000 per fish, and not $10,000 per human life. What I would suggest to you is that if you want to use cost-benefit analysis, let's look at all of the costs and all of the benefits, and don't be squeamish about the proposition that people have asthma and chronic bronchitis and other respiratory illnesses as a result of air pollution. And they may in fact have other non-respiratory diseases, if acid rain is resulting in heavy metal mobilization, especially mobilization of mercury and aluminum, as increasing evidence indicates.

The problem that I see is that efficiency is increasingly or too often injected at the wrong place, in deciding whether some action should be taken and not deciding what form the action should take. And that's one of the problems I have with the Waxman-Sikorski approach, although I understand completely why Henry Waxman is doing it--he's doing it because he considered acid rain to be a national problem and feels that people ought to contribute on a national basis. He thinks that it is a sufficiently serious problem to warrant enacting a law. To be perfectly blunt about it, he might not be able to say this but I can say it, he's trying to bribe people in the Middle West into accepting a pollution control program which they will otherwise bitterly resist, and might otherwise litigate to the point where, even if enacted, the law would become meaningless. That's why he's doing it and that's what I would say to my constituents if I were a representative of California or Hawaii or someplace else.

Finally, let me tell you what my conclusions are on all of this. One is, there is an overwhelming body of evidence to confirm what I would think that an enormous number of people intuitively believe--that pollution is bad for you. Pollution is intrinsically undesirable and what we should be doing is attempting to minimize pollution wherever we can. And this can be done efficiently; the Japanese have reduced emissions of sulfur dioxide, not concentrations as we have done in the United States, but total emissions of sulfur dioxide in a ten-year period by 50 percent. They are nonetheless whipping the pants off of us in the steel and automotive and every other industry and they are doing things like burning naptha which will generate electricity. It can be done and it can be done efficiently and it can be done and leave us competitive, both domestically and in the world-wide market. We should be minimizing pollution wherever we can, and that means going to lower sulfur fuels, going to coal washing if that's the most appropriate, it means going to flue gas desulfurization, if that is the best thing to do, and it means going to replacement capacity, if that's the most efficient thing to do. But you minimize pollution and you do it in the most efficient manner possible. At least, that's my judgment.

RESPONSE TO LESTER LAVE

Milton Russell

Assistant Administrator
Office of Policy Planning and Evaluation
United States Environmental Protection Agency
401 M Street, S.W.
Washington, DC 20460

I want to share with you my appreciation of the efforts of the organizers of this conference in bringing us all together to deal with a very important, complex and controversial issue: acid rain.

There are some key points that, while familiar, are worth restating because they are critical to deciding where we go from here.

1. Acid rain is not limited to "acid rain and the fish in the Adirondacks", but is the more general problem of acid deposition, with multiple causes and multiple effects.

2. The magnitude and specific nature of harmful effects from acid deposition are unclear, but probably significant.

3. Sulfur emissions have historically occurred at higher levels than at present; they might increase slightly in the near term, and then are likely to fall.

4. The relative contributions and harmful effects of sulfur to nitrogen oxide emissions are uncertain, but are likely to vary by time, location, and the health and ecological values to be protected.

5. At present it is uncertain whether damage to some systems is cumulative or if, instead, little additional damage will result unless emissions increase.

6. Remedial actions are going to take a long time to be implemented, and will cost a great deal.

7. The benefits of these actions will not be realized until sometime in the future.

The acid rain problem is thus quite different from that of an imminent hazard such as a toxic waste dump which is leaking into a drinking water supply. There we know what the problem is, how to fix it, and that it is urgent. We must act. With the acid deposition problem, we have time to consider our options before taking action. This difference is crucial. We have an opportunity to make sure that the actions taken are the most effective we can design.

This brings me to the points made by Lester Lave and Curtis Moore who, respectively, emphasize efficiency and equity. Efficiency is critical because, as a society, we simply do not have resources to waste in pursuing even as important a goal as environmental protection. It makes no sense to spend ten-billion dollars to correct a problem when six-billion will get the job done. But efficiency alone must not be the sole criterion. We also need to consider who pays the direct costs, and what ancillary effects actions have. This is the equity question. It has bedeviled those who have passed the threshold of deciding to act and seek a politically viable way to achieve emission reductions. The issue is confounded when the analysis of Ronald Jonash is factored in to illuminate the broad social impacts on communities which might lose jobs.

In my judgment we must broaden the discussion still further if we are to make wise environmental policy. Acid rain must be put into the context of the mandate of the Environmental Protection Agency. Our mission is to secure the healthiest and most ecologically sound environment for which the American people are willing to pay. To do that, we must reduce the worst risks first, because it is axiomatic that using resources against one problem means that fewer resources are available for something else.

Cost-benefit analysis is one of the tools that can help us determine what risks are worth reducing, and which ones to go after first. It serves to organize our thinking and to insure that all the effects are weighed as carefully as possible on the same scale. Lave makes an important point, however, in saying that cost-benefit analyses have not always included all of the factors that needed to be weighed, especially on the benefits side. This is a failure in implementation, not in principle.

I think it unlikely that analysts are, as Moore suggests, loath to include health care costs and health effects in their analysis. Nor, as Jonash suggests, are analysts unresponsive to second order effects such as those involved in infrastructure and disruption costs occasioned when actions lead to changes in the location of economic activity. Rather, analysts can fail by not pressing their work far enough, or by not specifying those unquantified effects that must be treated qualitatively. The latter sometimes include certain health effects, aesthetic considerations, and ecosystem stability, an effect particularly important to the acid rain issue. These unquantified impacts, and the uncertainties around them, should be identified for decisionmakers so that they may place their subjective values on these factors in final decisions. Bad cost-benefit analysis condemns the analysts, not the tool.

Given all of the environmental and health risks we need to address as a society, what should be our acid deposition control program? In confronting this question, we must first be mindful of our ignorance. Primarily because of uncertainties in the science, we don't know what an effective, much less an efficient, protective program involves. Until basic understanding is improved we don't know what emissions should be controlled, how best to control them, and thus how much an effective and efficient program would cost. We don't know what sources, in what places, should be controlled, or to what degree, and thus who would bear how much of the cost. (This leaves aside the thorny issue of whether that cost should be shifted to others, and if so, to whom). Further, we don't know what the benefits would be, and thus how good a buy control programs of different sizes and designs would be for the nation. With reference to the latter point, we must also look beyond the usual "acid rain" formulation to see what ancillary benefits (such as increased visibility) a reduction in sulfur loadings would bring. But let me emphasize, the conclusion to be drawn is not that we should necessarily forebear acting. Rather, as Lave quite correctly points out, decisions sometimes must be made without perfect, or sometimes even good, knowledge and information.

How, then, should we act in the face of large uncertainty? In my judgment, the decision process should begin by asking the question: "What are the environmental, economic, and social costs of being wrong, either in taking an action now or in waiting?" We can think of the answer in terms of our potential "regret function".

Pose the question this way: "How will we feel in ten years if we act now and are wrong?" Conversely, how will we feel in ten years if we wait, and that proves to have been wrong? Following upon the ten-year research program initiated by Congress in 1980, information has been accruing rapidly. What is the likely cost of waiting a year to make a decision? Two years? How much might we gain from the wiser decisions we could make using a larger information base? Responsible public policy involves not only designing strategy and implementation, but also the effective use of knowledge and information, and considering its rate of growth.

Any commitment to acid deposition reduction will absorb a significant portion of the resources available for pollution control. In deciding on a course of action we should keep in mind that what is involved is not a trade-off between money and pollution, but rather of one potential risk reduction against another. Being wrong in either direction has potential environmental consequences of great significance.

Decisions on acid rain will also have differential effects that raise the question of who pays and who benefits. Challenging equity issues transgress regional boundaries, span this generation and the next, and lie among industries, their employees and customers. Lave has suggested alternatives for resolving some of these equity issues, including the use of trading rights based on efficiency criteria. This idea has considerable appeal in that it "leaps over" administration and implementation complexities.

In conclusion, I would like to stress that the risk/risk trade-offs which we make as a society depend on three factors: (1) the information that can be brought to the table; (2) the methodology and judgment used in making the decision, and; (3) finally and most importantly, the values that society holds.

EPA has not been attempting to side-step a difficult decision. Rather, we have been grappling with these complex questions: "Is it worthwhile to wait? What do we need to know to make a wise decision? How can we get that information? What is efficient and effective action, if action is called for?" Answers are coming. Some people are satisfied now--on both sides of the issue. We are not. But we are pledged to reach a decision and take action when we are satisfied that we know enough, not just when we know everything.

RESPONSE TO LESTER LAVE: ECONOMIC IMPACTS OF ALTERNATIVE ACID RAIN CONTROL STRATEGIES

Ronald S. Jonash*

For Arthur D. Little, Inc. and Energy Ventures Analysis, Inc.
Ronald S. Jonash
Senior Consultant,
Economics and Development
Arthur D. Little, Inc.
Acorn Park
Cambridge, MA 02138

Numerous studies have been undertaken to determine the economic consequences of acid rain controls in the United States. Virtually all of the control programs evaluated thus far have focused on reducing SO_2 emissions at existing coal-fired power plants, and have differed only with respect to the magnitude and timing of SO_2 reduction. In addition, these studies have consistently focused almost exclusively on the cost to utilities and the resulting impact on the cost of electricity to consumers. Although recognizing that acid rain controls would lead to significant shifts in coal production and employment from high- to low-sulfur coal producing areas, none of these studies have attempted to estimate the indirect costs of these shifts in economic activity.

This paper summarizes the preliminary results of a study that examines the total economic consequences of acid rain controls, including the compliance cost to the utility industry, the impacts on those industries directly affected by utility compliance decisions, and the community adjustment costs associated with regional shifts in economic activity. The study was conducted by Arthur D. Little Inc., in conjunction with Energy Ventures Analysis, Inc., at the request of Consolidation Coal Company.

It should be noted that the study does not address the crucial economic and environmental issues of whether or not acid rain controls are cost justified or whether the programs currently under consideration will attain the desired environmental objectives. Clearly these questions must be answered first before addressing the secondary issue of how such controls should be implemented.

Table 1

Impact of Alternative Acid Rain Control Strategies on Total Compliance Costs (Present Value $1984 Billions)

	Fuel Switching Alternative	Technology-Based Controls	TBC vs. FSA Increasing (Decreasing)
Industry Adjustment Costs			
Electric Utility Costs	$32.0	$35.7	$3.7
Mine Closure Costs	1.3	0.4	(0.9)
Subtotal	33.3	36.1	2.8
Community Adjustment Costs			
Household/Worker Costs	1.2	0.3	(0.9)
Private/Public Development Costs	9.9	2.6	(7.3)
Subtotal	11.1	2.9	(8.2)
TOTAL COMPLIANCE COSTS	$44.4	$39.0	($5.4)

A comparison of the impacts of two alternative acid rain control strategies, one emphasizing fuel switching, the other technology-based controls, indicates that total compliance costs are likely to be lower under a technology-based control program. The direct compliance costs to electric utilities would be somewhat higher under technology-based controls, but these higher costs would be more than offset by lower indirect costs associated with shifts in economic activity. Technology-based controls would have less of an adverse effect on the mining industry, on shifts in employment and population migration, on worker productivity and training costs, and on public and private community investment requirements.

As summarized in Table 1, the total present value of compliance costs would be approximately $39 billion under technology-based controls as compared with $44 billion under fuel switching.[1]

The study also found that community fiscal conditions are likely to be less adversely affected under a technology-based control program. The U.S. balance of trade is also likely to be stronger under a technology-based control program.

[1]All cost estimates in this study are in 1984 dollars.

ALTERNATIVE COMPLIANCE STRATEGIES

Each of the acid rain control programs evaluated in this study was modeled along the lines of previously proposed legislation and was designed to achieve the same 8 million ton per year reduction in SO_2 emissions by 1995. In each case, approximately 0.8 million tons of SO_2 reduction would be attained through compliance with state implementation plans and reduced coal consumption. The remainder of the required reductions was allocated to states and utilities in the eastern U.S. in proportion to their share of 1980 utility emissions in excess of 1.5 pounds of SO_2 per million BTU. Limited interstate trading was permitted for individual utilities. The principal differences between the alternative strategies were:

- Under the Fuel Switching Alternative (FSA), utilities were assumed to achieve their required reductions in SO_2 emissions by switching to lower sulfur fuels unless the cost of retrofitting flue gas desulfurization systems (scrubbers) was at least 25 percent less than the cost of switching to a lower sulfur fuel. The 25 percent "hurdle rate" for retrofitting scrubbers was included to reflect increasing utility resistance to large capital investments and other intangible impediments to the retrofitting of scrubbers.

- Under Technology Based Controls (TBC), utilities were assumed to achieve their required reductions in SO_2 emissions by retrofitting scrubbers unless scrubbing was not technically feasible or involved costs at least 25 percent higher than the costs of fuel switching.

DIRECT INDUSTRY IMPACTS

Direct industry impacts of acid rain controls consist primarily of utility fuel use and pollution control investment decisions, coal production shifts and associated mine development and closures, and coal transportation shifts. Table 2 indicates the methods of SO_2 reduction under each of the acid rain control strategies and the impacts on other industry operations.

- Electric Utilities. Under FSA, 70 percent of the required reduction in SO_2 emissions would be achieved by fuel switching and 30 percent by scrubbing. Under TBC, 35 percent of the required reduction would be achieved by fuel switching and 65 percent by scrubbing.

- Coal Mining. Total coal production in the United States was assumed to remain essentially the same under both control strategies. However, there would be a shift from high- to low-sulfur coal producing areas of 110 million tons per year under FSA versus a shift of 37 million tons per year under TBC.

- Coal Transportation. Under FSA and TBC, the increase in coal haulage would be 70 billion and 39 billion ton-miles per year, respectively. Most of this increase is in railroad transportation, particularly in the West.

- Pollution Control Equipment. The pollution control equipment industry would increase investment for retrofitting scrubbers (19,000 MW of capacity under FSA, 49,000 MW under TBC), and for modifying ESP systems (86,000 MW under FSA, 40,000 MW under TBC).

Table 2

Impact of Alternative Acid Rain Control Strategies on Industry Operations

	Fuel Switching Alternative	Technology-Based Controls	TBC vs. FSA Increasing (Decreasing)
Utility SO_2 Reduction (MMTPY)			
Fuel Switching	5.1	2.6	(2.5)
Scrubbing	2.1	4.7	2.6
Other[1]	0.8	0.7	(0.1)
TOTAL	8.0	8.0	0.0
Pollution Control Equipment (MW of Generating Capacity)			
FGD Scrubber Retrofits	19,000	49,000	30,000
ESP Modifications	86,000	40,000	(46,000)
Coal Production (MMTPY)			
Northern Appalachia	(47.4)	(12.3)	35.1
Illinois Basin/Interior	(62.3)	(24.7)	37.6
Central/Southern Appalachia	76.2	15.4	(60.8)
West	38.5	28.1	(10.4)
TOTAL	5.0	6.5	1.5
Coal Transportation (Billion Ton Miles)			
Eastern Rail	10.1	1.2	(8.9)
Western Rail	48.7	33.7	(15.0)
Barge	11.5	3.6	(7.9)
TOTAL	70.3	38.5	(31.8)

[1]Includes reductions in SO_2 resulting from compliance with state implementation plans for SO_2 and reduced coal consumption within certain utility systems.

INDUSTRY COMPLIANCE COSTS

Industry compliance costs consist primarily of the direct costs of utility compliance, including the increased costs for fuel, transportation and pollution control equipment. Except for high-sulfur mine closure costs, other industry compliance costs are assumed to be included in the fuel, transportation and pollution control costs paid by electric utilities. Table 3 shows the breakdown of industry costs of compliance for the alternative strategies.

Table 3

Impact of Alternative Acid Rain Control Strategies on Industry Compliance Costs ($1984 Millions)

	Fuel Switching Alternative	Technology-Based Controls	TBC vs. FSA Increasing (Decreasing)
Utility Capital Costs for Pollution Control Equipment			
FGD Scrubbers	$4,840	$12,440	$7,600
ESP Modifications	930	490	(440)
TOTAL	5,770	12,930	7,160
ANNUALIZED CAPITAL COSTS	553	1,185	632
Utility Operating Costs			
Scrubber Operations	320	807	487
Fuel	376	82	(294)
Transportation	1,091	549	(542)
ANNUALIZED OPERATING COSTS	1,787	1,438	(349)
Total Utility Compliance Costs			
Annualized Costs	2,340	2,623	283
Present Value	31,961	35,709	3,748
Premature Mine Abandonment Costs			
Present Value	1,291	404	(887)
TOTAL PRESENT VALUE OF INDUSTRY COMPLIANCE COSTS	$33,252	$36,113	$2,861

- Electric Utilities. The present value cost of utility compliance would be $32.0 billion under FSA as compared with $35.7 billion under TBC. The resulting increase in average utility revenues required in the 31-state control area would be about 2.1 percent under FSA and 2.3 percent under TBC. Without regional cost spreading mechanisms, utility systems in some midwestern states would have to increase their utility rates by as much as 20 percent. For the purposes of this study, we have not assessed the regional impacts on utility rates because of the potential for cost spreading mechanisms and because of the uncertainties of individual utility rate treatment under state public utility commissions.

- Coal Mining. Capital costs for additional low-sulfur coal production capacity would be approximately $9.7 billion under FSA as compared with $4.2 billion under TBC. The capital cost requirements for new low-sulfur production capacity are assumed to be internalized in low-sulfur coal prices. However, mine abandonment costs for the premature closing of existing high-sulfur coal mines are an externality which must be added to the cost of complying with acid rain controls. These costs are estimated to be $1.3 billion under FSA as compared to $0.4 billion under TBC.

- Pollution Control Equipment. Capital expenditures for pollution control equipment (primarily scrubber retrofits and ESP modifications) would total $5.8 billion under FSA and $12.9 billion under TBC.

- Coal Transportation. Transportation costs would increase substantially under both acid rain control strategies because of the displacement of locally produced high-sulfur coals by low-sulfur coals, which must be transported over longer distances. The additional capital investment requirements for rail and river transportation equipment would be approximately $0.8 billion under FSA as compared with $0.4 billion under TBC.

EMPLOYMENT/HOUSEHOLD MIGRATION IMPACTS

While both strategies are estimated to result in approximately the same number of jobs nationally, the technology-based controls would result in substantially less job and household displacement impacts. A geographical breakdown of employment impacts under the alternative strategies is provided in Table 4.

- National Employment. National employment is estimated to increase by approximately the same number of jobs under both acid rain control strategies, 30,000 under FSA and 31,000 under TBC. National coal mining employment remains about the same under both strategies with the principal increases occurring nationally in utilities (10,500 under TBC, 3,600 under FSA), transportation (7,200 under TBC, 13,700 under FSA), and trade and services (10,000 under TBC, 7,300 under FSA).

Table 4

Impact of Alternative Acid Rain Control
Strategies on Total Employment Versus 1995 Base Case

	Fuel Switching Alternative			Technology-Based Controls		
	Mining	Other	Total	Mining	Other	Total
Coal Producing Regions of Major Coal Mining States						
Northern Appalachia	(15,700)	(20,000)	(35,700)	(4,100)	(1,600)	(5,700)
Illinois Basin/Interior	(16,100)	(22,300)	(38,400)	(6,100)	(4,400	(10,500)
Central/Southern Appalachia	30,400	36,000	66,400	6,400	7,600	14,000
West	2,000	4,600	6,600	1,400	3,300	4,700
TOTAL MAJOR COAL PRODUCING REGIONS	600	(1,700)	(1,100)	(2,400)	4,900	2,500
Rest of U.S.	(500)	31,300	30,800	(500)	28,800	28,300
NATIONAL TOTAL	100	29,600	29,700	(2,900)	33,700	30,800

- Employment Shifts. Employment shifts, however, are estimated to be about four times greater under FSA than under TBC, 58,000 jobs under FSA compared with 16,000 under TBC. About half of these shifts represent coal mining jobs that would be lost in Northern Appalachia and the Illinois Basin and recreated in Central Appalachia and the West. The balance of employment shifts under the two strategies represents the net effect of increases in utility, pollution control, and coal transportation employment, partially offset by decreases in service and related sectors. Most of the job declines are estimated to take place in communities that are stable or declining with little growth potential to absorb displaced workers, particularly coal miners. Most of the job increases are estimated to take place in low-sulfur coal mining areas that already are projected to be growing rapidly in the absence of acid rain controls.

- Household Migration Impacts. Household migration impacts are also projected to be substantial due to the distribution of job displacement effects. Households which would have to move out of high-sulfur coal producing areas are estimated to total 51,000 under FSA and 12,000 under TBC. Households which would have to move into low-sulfur coal producing areas are estimated to total 61,000 under FSA and 15,000 under TBC.

COMMUNITY ADJUSTMENT COSTS

Community adjustment costs include household and worker adjustment costs and public and private community investment requirements resulting from development shifts in both growing and declining communities. Table 5 provides a breakdown of the present value of these costs, which total $11.1 billion under FSA and $2.9 billion under TBC. Since most high-sulfur coal producing areas already have surplus private and public infrastructure, investment savings due to further population declines would be relatively low. On the other hand, most low-sulfur coal producing areas are expected to be rapidly growing and capacity-constrained by the early 1990's. As a result, low-sulfur coal producing areas would require the full increment of community investment in order to accommodate additional growth.

- Worker Adjustment Costs. Worker adjustment costs include the costs of training as well as reduced productivity. Training costs are estimated to have a present value of $410 million under FSA versus $110 million under TBC. For the purpose of this study, unemployment and social welfare benefit payments were used as a conservative proxy for losses in worker productivity. These payments are estimated to amount to $520 million under FSA and $250 million under TBC.

- Household Adjustment Costs. Household adjustment costs in this analysis were limited to the direct moving costs associated with additional household migration ($310 million under FSA and $80 million under TBC). These cost estimates exclude the wealth effects of households having to sell low valued homes in declining areas and having to purchase higher valued homes in growing areas. They also exclude the psychological and health costs associated with the loss of employment and involuntary moves.

- Private Development Costs. Private development costs include the private investment requirements for housing and commercial development in low-sulfur areas offset by the reduced private investment requirements in high-sulfur areas. These costs have an estimated present value of $7.3 billion under FSA and $1.9 billion under TBC.

- Public Infrastructure Costs. Public infrastructure costs represent the public sector complement to the private community investment costs summarized above. They include water, sewer, roads, parks, schools, and other government facilities and have an estimated present value of $2.6 billion under FSA and $660 million under TBC.

Table 5

Impact of Alternative Acid Rain Control Strategies on Community Adjustment Compliance Costs (Present Value $1984 Millions)

	Fuel Switching Alternative	Technology-Based Controls	TBC vs. FSA Increasing (Decreasing)
Household/Worker Adjustment Costs			
Household Moving Costs	$310	$80	$(230)
Worker Training Costs	410	110	(300)
Worker Productivity Losses[1]	520	150	(370)
Subtotal	1,240	340	(900)
Community Development Costs			
Private Development Costs	7,270	1,870	(5,400)
Public Infrastructure Costs	2,570	660	(1,910)
Subtotal	9,840	2,530	(7,310)
TOTAL COMMUNITY ADJUSTMENT COSTS	$11,080	$2,870	$(8,210)

[1]Unemployment/Welfare transfers used as proxy for this cost.

OTHER IMPACTS

In addition to the impacts and adjustment costs previously described, the two strategies are likely to have significantly different impacts on community fiscal conditions and the U.S. balance of trade.

- Net Community Fiscal Deficits are estimated to be lower under TBC in both high- and low-sulfur areas. In growing low-sulfur communities, increases in severance taxes and other revenues are estimated to occur faster than increases in operating costs. However, operating surpluses would not be sufficient under either strategy to offset public investment requirements, resulting in a net $250-million deficit under TBC and a $1.5-billion deficit under FSA. In declining high-sulfur communities, decreases in severance taxes and other revenues are estimated to occur faster than decreases in operating costs. Consequently, there would be a fiscal operating deficit in these areas. Despite potential savings in avoided replacement costs for public infrastructure, the net effect is still a fiscal deficit in high-sulfur areas of an estimated $880 million under FSA and $250 million under TBC.

- U.S. Balance of Trade is affected under the alternative strategies as a result of altered oil import requirements and altered low-sulfur coal exports. The technology-based controls alternative is likely to result in a smaller estimated increase in oil imports ($50 million/year compared with $110 million/year) and a smaller estimated decrease in coal exports. If only 10 percent of the additional 70 million tons/year of domestic low-sulfur coal demand under FSA were to displace exports, the value of these lost exports would be in excess of $400 million/year.

In summary, the decision concerning which of many acid rain compliance strategies is in the best interest of the nation will involve a complicated weighing of many, often conflicting factors. While the study does not attempt to reach a final judgment, it does provide new information on the critical elements involved in such a judgment. The major conclusion of the study is that utility compliance costs are only part of the total costs of acid rain controls and the indirect economic and social impacts of such programs, which are substantial, must be given equal consideration before federal or state actions are taken on this issue.

DISCUSSION

IAN M. TORRENS (OECD): There sometimes tends to be a suspicion that cost-benefit analysis is an alternative to policy action. I was interested to hear Curtis describe his impressions of his recent trip to Japan. In fact, Japan does not, to my knowledge, rely to any great extent on sophisticated cost-benefit analysis in the environment field. In the area of air pollution, for example, the government sets limits to pollutant emissions which owe their origins to general air quality standards. Local authorities can and often do set stiffer emission limits, particularly in industrialized urban areas. Then the industries or utilities get together with local authorities and negotiate specific emission limits for particular installations which tend to be worked out on the basis of what pollution control technology now available can achieve. These local authorities are often representing the people who have to pay the costs of pollution control (e.g. via electricity prices from the utility), and they have no doubt themselves as to the perceived cost-benefit balance. Finally, having agreed on a negotiated set of emission limits, the company or utility gets its engineers to work on meeting these limits at the lowest possible cost. In fact, what usually happens is that they succeed in doing much better; one power plant I have seen emits half as much SO_2 in reality as its negotiated agreement with the local authority would permit. Sophisticated cost-benefit analysis plays no significant role in this whole process.

At the end of the great American debate on air pollution/acid rain control, if and when government and legislators finally act, it might be interesting to apply a little of the "ex-post analysis" we heard suggested yesterday, and work out what the total debate has cost, all the lawyers' and consultants' fees, the R&D spending, the legislative costs, and so on and so forth. I suspect it would be up there in the billion dollar range without having achieved many benefits in the form of reduced air pollution while it was in progress!

CHARLOTTE ZIEVE (University of Wisconsin): Mr. Lave, I am concerned when we don't look at the system. When you talk about urging coal washing I would like to know if you thought about what we do in terms of the water supply and disposal of whatever you've done when you wash the coal. I think that a lot of times we talk about solutions without considering the total picture; as a matter of fact I've heard it said that acid deposition is a result of not looking down the road. The second point is directed to Milton Russell; I'd like to ask him if he thinks the next conference should be on risk assessment.

LESTER LAVE: Your point is extremely well taken. Much of today's acid deposition comes from "cheap" solutions in the past. Tall stacks were the cheapest way to solve local sulfur dioxide problems. However, they exported pollution to exacerbate acid deposition downwind. This time we have to be careful that coal washing residuals do not cause new environmental problems.

MILTON RUSSELL: I think risk assessment is an important point in this case. We do have to take into account what these risks are. The fact is that these are issues that are not going to be resolved simply. It is easy enough to enunciate simple solutions but much more difficult to try to think through what the longer-term and broader implications are, not only for air, not only for sulfur acids, but broader societal economic impacts such as we have been discussing today.

ROBERT FENTON (Winnipeg, Canada): If I recall Professor Crocker's discussion yesterday, he was talking about five billion dollars a year in benefits and, if my calculations are correct, into perpetuity that has a capital value of fifty billion dollars. Based on the Arthur Little presentation this morning, it looked like the discussion about a present value of about fifty billion dollars in costs, so it sounds to me as if we have an interesting area of discussion here, and I think that is something that should be followed up. My question is: I heard two speakers about cost impacts this morning; one stated that the acid rain control process would be very expensive and the other speaker spoke about a one percent increase in utility costs in Illinois. Is the Illinois case representative of the costs that would be incurred? How do these costs work out in terms of possible kilowatt hours and costs per capita? Are we talking about costs per person that are such that people feel that the costs will not significantly bias decisions one way or another?

LESTER LAVE: In regard to the first point, it certainly is true that we ought to be looking at social benefits and costs. The best social benefit analysis that I can do is biased towards not doing anything; there are many effects that I can't quantify and have ignored. This "best" analysis nonetheless indicates that quite substantial abatement of sulfur dioxide from stationary sources is worthwhile. Abatement should be much more stringent than current effort. Not everyone agrees with this analysis; particularly there is criticism of the estimates of the health effects of sulfur oxides. There is a lot of uncertainty there. I think I am right, but I'll concede that they have reasonable objections.

The point is that many current estimates of health effects are large enough to justify stringent abatement by themselves. Thus we ought to concentrate on resolving some of these uncertainties.

In regard to the second point, no state is representative of the whole country; they're all going to be different; probably the highest costs are going to be in a state like Ohio. However the cost of an efficient abatement program is affordable. My personal opinion is that we ought to proceed immediately with the low-cost abatement method such as coal preparation and some low-sulfur coal. The more expensive methods could await the resolution of uncertainties, such as health effects. I probably want more abatement now than some others, but at least we can all agree to get started immediately on the cheap measures.

QUESTIONER (Indiana): First, on the not unreasonable issue of health effects. I want to point out that we do still have the Clean Air Act for air quality standards to deal with potential health effects of acid rain. Second, in regard to the costs and Mr. Lave's point that Illinois is an example. We are looking at state averages there, thus the Commonwealth Edison system is not affected very much under the Waxman-Sikorski bill. If you are looking at impacts in Southern Illinois, they are much greater, and these are significant impacts on individuals whose rates are being raised.

In Indiana, looking at our current revenues with no growth in costs, the increase to our customers with scrubbers would be 48 percent.

LESTER LAVE: Have you worked out what Mitchell-Stafford would cost?

QUESTIONER: Yes, we have and that would run around 38 percent.

HERBERT VISSCHER (Georgia): This question is directed to Mr. Jonash. We saw a less than optimum set of Clean Air Act amendments in 1978, adopted out of concern for unemployment of high-sulfur coal miners and now we see recommendations on the acid rain deposition issue that are affected by unemployment in high-sulfur coal mining. My question is, are you proposing that all significant areas of technological unemployment become an area of major concern, or a higher concern level than we have seen in the past? Another example might be unemployed auto workers displaced by higher technology in the past. Or is this a single incident of technological unemployment to be given major attention in federal policy?

RONALD JONASH: I think that technological unemployment in your home state, Texas and other states also deserves attention and our presentation was not meant to say that the government needs to step in and save jobs or save dying industries in cases of technological unemployment. Where we are proposing a regulatory scenario which will create technological unemployment, we should strive to come up with the most efficient or cost-effective regulatory scenario. We are not trying to save high-sulfur coal mining jobs per se; what we are trying to determine is what is the least cost to society. That cost includes some costs borne by coal miners, some costs borne by communities and some costs borne by utilities. Another way to look at it is as follows: if the utilities or government were to pay the full costs of the disruption, you would have a completely different decision by the utilities as to what the least-cost alternative would be. For example, if the utilities in Illinois would be making the compliance decision on the basis of not only what the least cost is to them, but what the least cost is to the State of Illinois in terms of community effects in Southern Illinois, you might have a different decision than if the utilities were to make decisions on the basis of what has been euphemistically called "freedom of choice." The bottom line is that regulation-induced unemployment is a different situation than normal technological unemployment in steel, automobile, textile and other industries.

CURTIS MOORE: I want to comment on Lester Lave's point about waiting. People talk about waiting one year, two years, or three years, and I would observe that it was in May of 1979 that Sen. Stafford, with me staffing him, went to the floor of the U.S. Senate and first sought an acid rain control bill. I would also like to observe that what we're doing here is re-arguing a decision that was made in 1970. This country decided in 1970 that we were going to reduce air pollution. In Japan they reduced air pollution by 50 percent. In the United States, instead of reducing air pollution, we changed its form by adopting a least-cost approach to pollution control. We know what the health effects of sulfur dioxide are; we know that sulfur dioxide causes respiratory illness; there is no question about that. We are now arguing about what the health effects of sulfates are. We are arguing about the health effects of sulfates in large measure because we avoided the mandate of the 1970 law. I might observe too that we would be much further along right now if the state of Ohio had adopted a state implementation plan in compliance with the Clean Air Act, if the state of

Indiana had adopted a SO_2 state implementation plan (SIP) in compliance with the Clean Air Act, and if the state of Illinois had not had the implementation of its SO_2 SIP delayed by ten years of litigation. If those states and the other states that are the major producers of sulfur dioxide in this country had complied with the intent of the Clean Air Act and reduced emissions, rather than seeking what I hear so many people say, time and time again, and that is "least cost", we would not now be arguing about whether there are health effects or whether there are forest effects, or whether there are other effects from acid rain. And from the point of view of a state like Vermont where 75 percent of the red spruce, in an absolutely pristine area that you can only reach by long hiking, have been destroyed, and where the only viable hypothesis so far is acid precipitation, where the decline in sugar maple production is on the order of 40 percent, that from the perspective of the state of Vermont, if somebody is going to suffer a rate increase of 10 or 20 or 30 percent because of acid rain controls, it is because they have had 14 years of rate decreases of 15 or 20 or 30 percent, compared with states like Vermont which have complied with the intent of the Clean Air Act and reduced pollution. Excuse me if I got a little forceful there, but I think my Senator would want me to say that.

RISK ASSESSMENT AND DECISION CRITERIA

CAN RISK-BENEFIT ANALYSIS BE USED IN RESOLVING THE ACID RAIN PROBLEM?

Richard Wilson

Professor of Physics, Department Chairman
Energy and Environmental Policy Center
Harvard University
Cambridge, MA 02138

INTRODUCTION

Risk analysis can be described as the use of common sense combined with arithmetic. Computers can ensure that the arithmetic, even complex arithmetic,is done right. However common sense is not as common as it should be. The errors in a risk-benefit analysis are not usually errors of arithmetic, but errors of common sense.

I will explain this, with reference to the problems of acid rain, by a parody of a risk-benefit analysis using a simple recipe that an economist might suggest.

I start by noting that the financial benefits*, B(t) dt, obtaining between the times t and t + dt can be calculated. So also can the financial costs C(t) dt. We can also estimate the risks R(t) dt which accrue at these times. Then the Net Present Value equation can be used to decide whether or not to proceed with the project under discussion.

$$NPV = \int_t \frac{B(t) - C(t) - \alpha R(t)}{(1 + i)^t} \, dt \qquad \text{(Eq. 1)}$$

i is the interest rate or discount rate and α is a parameter to transform the risk into financial units.

*Editor's Note. The definition of benefits and costs as used by Wilson is different from that in the table of contents and other articles in this book. In Wilson's article, benefits are the financial benefits of continuing the polluting activity, costs are the financial costs of pollution control, and risks are the potential harms to human health. In the rest of the contents, costs refer to the costs of controlling pollution, and benefits refer to pollution damages avoided, i.e. αR(t) with the opposite sign. It is assumed that the activity is going ahead, so that the trade-off is between C(t) and αR(t). Wilson's analysis thus poses the question--shall we proceed with project X, while the other analyses in the book are predicated on the question--shall we require control strategy X.

According to this naive recipe, if NPV > 0 the benefits exceed risks and we should proceed with the project. The parameter α, which looks so small in the equation, can "obviously" be determined and once determined it is obviously a constant, and can be used for all decisions.

I reject this simple approach, not because the equation is wrong, but because I have cast it in a misleading form. In particular I propose to show that α is not a constant, and that a discussion of its value is central to a risk assessment and the decision process the assessment hopes to illuminate. I also propose to show that the uncertainties in determining R(t) are crucial.

I first do this, not with the acid rain problem, but with a simpler, but related problem: the mortality due to pollution of air by acid sulfates and other particulates. I do this because a lot more is known about the problem scientifically and the public perception is also better understood. Moreover, I have been co-author of a book about it (Wilson et al., 1980). Below I will relate the changes in the argument that may be necessary as acid rain is discussed.

THE COST TO REDUCE A RISK

The innocuous looking parameter α has the dimensions "dollars per life". It has been described by opponents of risk analyses as an attempt to find the "value of a life." That is an unnecessary and undesirable interpretation of the algebra and arithmetic, undesirable because it introduces theological rather than practical questions. But it does warn us that α might not be considered a constant. Let me instead consider it as the "cost per fatality averted." Cohen (1980) prepared a table of several instances where society spends sums to avert fatalities. Table I is a modification of this by Crouch and Wilson (1982). Others have prepared similar tables, but I prefer that of Cohen, because Cohen went through most of the original data, correcting for omissions and making the estimates consistent.

It is evident that α is *not* a constant. Society could save lives, at $100 a head, by expanding immunization in Indonesia; society does attempt to save lives by spending $30 million per head on safety in US coal mines. α varies nearly a factor of 10 million! Furthermore even though most members of society do not know these numbers in detail, it is evident that the general spread is known. There is no pressure on Congress to appropriate money for immunization in Indonesia; there is some pressure to spend more on coal mine safety in general.

The numbers can, I believe, be grouped. For saving lives in the Third World we spend very little, whatever our pious declarations may be. For medical life-saving activities such as radiotherapy, we spend about $50,000 per life; yet the cost of kidney dialysis is sufficiently large, $200,000, that questions have been publicly raised about this activity. We spend $1,000,000 per life and more on EPA preventive activities. This is roughly in accord with the maxim, "an ounce of prevention is worth a pound of cure." Coal mining poses a very large risk to those employed in the mines, and we will all be happy to reduce this large risk even though it is expensive in terms of cost per fatality averted.

Thus I come to the conclusion that although α is not a constant, the formulation can still be a useful guide to the decisionmaker, provided he can relate the proposed activity to one of the activities in the groups of Table I.

A very bad example of misuse of risk assessment was presented this last summer by EPA. In a discussion of proposed changes in the particulate standard, they used a constant value of α. Worse still they did not state clearly what value they used, and it was left up to the reader to determine it from their conclusions. Their value, α = \$300,000 per fatality averted, may be reasonable, but it should be argued. By this omission, their whole report becomes useless as described by Wilson, Crouch, et al. (1984).

The formalism can be modified for environmental decisions, as distinct from public health decisions, by altering the meaning of R. There are a myriad of environmental problems, but I will pick one, the eutrophication and ultimate destruction of lakes by acid rain. R(t) dt might be the risk of destruction of a lake and α the "cost of lake reprieved." α is still not constant; nor is the cost linear with the number of lakes. If there are 1000 lakes, many of us are willing to see 10 destroyed or covered by a parking lot, but not all 1000. The last one we will guard jealously, even with our lives. This nonlinearity is not obvious in the case of public health decisions.

Persons differ widely in their approaches to the environment. α can thus vary widely even over the small group in this room. But there is not such a wide variation between distant environmental objects and local ones. The world applauded the efforts of Soviet environmentalists to save Lake Baykal and there is almost as much interest in the U.S. in saving a rhinoceros in the Serengeti Plain of Tanzania and Kenya as saving the bald eagle closer to home.

UNCERTAINTIES IN THE RISK CALCULATION

The calculation of the risk R(t) dt is often difficult. Again I illustrate using the estimates of health effects of air pollution.

The data is voluminous, but not much of it is direct. The most direct data are in Figure 1 which shows the death rate in London week by week in December 1952 (solid line) and compares it to an average over other years (dashed line). No one to whom I have talked doubts that the difference in area under the two curves, 2900 excess deaths, had a cause, and when the further information is provided that during the first week in December there was a bad London fog, an air pollution incident, everyone attributes these deaths to air pollution.

The level of air pollution was high, about 4.5 milligram per m^3. The problem is to estimate the excess deaths, if any, at today's air pollution levels in the U.S., which are about 50 $\mu g/m^3$ over the eastern U.S. I can make some sweeping assumptions, each one of which produces a huge uncertainty. First, I make a linearity assumption, that reducing the concentration of pollutants by a factor of 90 reduces the death rate by a factor of 90 and by a factor of 90 only. Second, I assume that the relevant exposure is a long-term average and that exposure over 52 weeks is 26 times worse

Table 1

Cost Per Fatality Averted (1975 Dollars) Implied by Various Societal Activities

Item	Fatality Averted	Item	Fatality Averted
Medical Screening and Care		Miscellaneous Non-Radiation	
Cervical cancer	25,000	Expanded immunization in Indonesia	100
Breast Cancer	80,000	Food for overseas relief	5,300
Lung cancer	70,000	Sulfur scrubbers in power plants	500,000
Colorectal cancer:		Smoke alarms in homes	250,000
Fecal blood tests	10,000	Higher pay for risky jobs	260,000
Proctoscopy	30,000	Coal mine safety	22,000,000
Multiple screening	26,000	Other mine safety	34,000,000
Hypertension control	75,000	Coke fume standards	4,500,000
Kidney dialysis	200,000	Air Force pilot safety	2,000,000
Mobile intensive care units	30,000	Civilian aircraft (France)	1,200,000
Traffic Safety		Radiation Related Activities	
Auto safety equipment, 1966-70	130,000	Radium in drinking water	2,500,000
Steering column improvement	100,000	Medical X-ray equipment	3,600
Air bags (driver only)	320,000	ICRP recommendations	320,000
Tire inspection	400,000	OMB guidelines	7,000,000
Rescue helicopters	65,000	Radwaste practice, general	10,000,000
Passive 3-point harness	250,000	Radwaste practice -- ^{131}I	100,000,000
Passive torso belt-knee bar	110,000	Defense high level waste	200,000,000
Driver education	90,000	Civilian high level waste	
Highway Construction -- maintenance practice	20,000	No discounting	18,000,000
Regulatory and warning signs	34,000	Discounting (1%/year)	1,000,000,000
Guardrail Improvements	34,000		
Skid resistance	42,000		
Bridge rails and parapets	46,000		
Wrong way entry avoidance	50,000		
Impact absorbing roadside devices	108,000		
Breakaway sign, lighting posts	116,000		
Median barrier improvement	228,000		
Clear roadside recovery area	284,000	Source: Cohen (1980).	

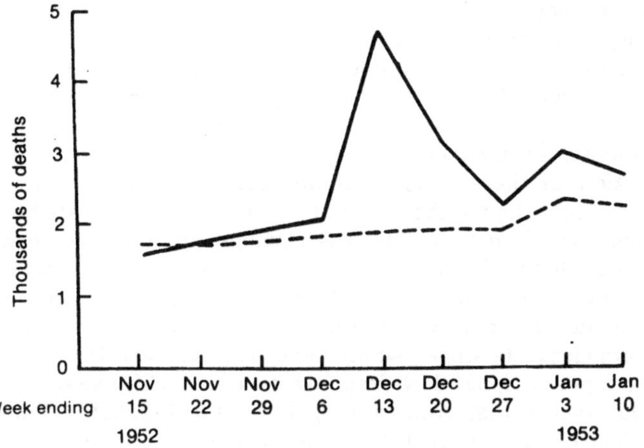

Figure 1. Deaths in Air Pollution Episode

Deaths registered in greater London associated with air pollution episode of December 5-8, 1952. Broken line indicates average deaths. (From J.R. Goldsmith and L.T. Friberg, "Effects on Human Health," in A. Stern (ed), *Air Pollution*, New York: Academic Press, 1977.)

than exposure over 2 weeks. Third, I assume that the 150 million Americans in the eastern U.S. respond to air pollution like the 5 million Londoners. The yearly death rate comes out to be:

$$3900 \times \frac{50\ \mu g/m^3}{4.5\ mg/m^3} \times \frac{52\ weeks}{2\ weeks} \times \frac{150\ million}{5\ million} \approx 33{,}800$$

This is a large number. Can we prove it? Not rigorously. Can we disprove it? Not rigorously, because it is only 2 percent of the death rate and it is spread out over a large population. Other evidence can be brought to bear which follows earlier work of Lave and Seskin (1970), and which is discussed in Chapter VI of Wilson et al. (1980) and a more recent paper (Evans et al., 1984). These authors find a correlation between mortality and air pollution indicators in the U.S. When multiplied out, their correlation factor naively suggests about 50,000 annual deaths at today's levels. Although Evans et al. point out that most public health experts do not believe that the correlation is causal, they also state that Lave and Seskin do believe that it is causal.

Some experts on public health who should know better ignore the possibility that the naive calculations above, and the more sophisticated ones of Lave and Seskin and of Evans et al. might be right. Then since they see no risk, benefit exceeds risk and there is no air pollution problem. This I characterize as an OMB risk-benefit calculation. I object to these not only because they are misleading or even wrong, but because they misuse the English language without following Humpty Dumpty and paying the words, and the listeners of the words, extra (Carroll, 1870).

The problem is a typical one in risk assessment. It arises because either there is an extrapolation, or there is doubt about causality. The reason that many air pollution experts doubt the numbers is that for other, often unstated but nevertheless sensible reasons, they do not consider the numbers to be in accord with the rest of their scientific knowledge. Why should there be a correlation of mortality with sulfur dioxide? SO_2 itself seems relatively innocuous and nonirritant in most tests. If the particulates are the problem, why are air pollution particulates much worse than particulates from cigarette smoke whose concentrations are 100 times higher? Why should there be a proportional dose-response relation? Fifty years ago it was widely believed that for everything, even cancer, there is a threshold. This is less widely believed now. A risk assessor must take differences of opinion into account.

In addressing this question I note that the risk assessor, like the scientist, must have imagination. He must use imaginative models. For example, someone who has been exposed to air pollution, whether industrial or cigarette smoke, is likely to have reduced lung function. This reduced lung function might, 30 years later, make him more susceptible to early death from some seemingly unrelated cause such as falling downstairs (Evans, 1982).

This tendency to include in the list of risks which I call $\Sigma R(t)$ only those which are well known ignores the elementary fact that the word "risk" itself is an expression of uncertainty. I have heard colleagues say that "we should only regulate certain risks" (meaning risks that are certain, not particular risks). Then they do not discuss matters on which there is a difference of opinion. Such a scientist is mixing his scientific statements and his policy statement. If you trust him, well and good; but he does not help a decisionmaker in his attempts to understand the complexities of the decision.

What do we do in light of this disagreement and hence uncertainty? This is a crucial question for the making of the decision, should not be sloughed over by sloppy analysis and certainly should not be ignored. In an excellent paper, Morgan et al. (1984) discuss ways of addressing uncertainty. In a number of papers in other areas of risk, Crouch and Wilson (1981) and Crouch, Wilson and Zeise (1983) have argued that the risk assessor should calculate a probability distribution of the risk coefficient, or in particular at least two moments thereof, the median and the upper 98th percentile.

When we do this, it becomes clear that one of the questions for decision is how careful to be. In general, for decisions before a pollutant is allowed, we might take the 98th percentile; but it is not so obvious when billions of dollars are necessary to reduce the risk. However, this is not for the risk assessor to decide, only to discuss and hopefully to illuminate. I do not here show how to calculate the risks of destruction of a lake; this is still very uncertain. An ardent environmentalist might argue that we must be extraordinarily careful to calculate on the "safe" side or, in analytical language, to use the upper 98th percentile of the probability for a large value of α (the value of a lake). Then we should spend an unlimited amount of money on reducing acid rain.

The assumption of linearity I made for the relation of health effects and the exposure (the dose-response relation) is far less likely to hold for environmental risks. There are many reasons for this, some of which are based upon the science of eutrophication itself. I believe a more important question is the nonlinearity of one's values. We all accept that some U.S. wilderness has given way to civilization, to build our cities. But to destroy the last wilderness would bring out environmentalists in droves. Similarly we are usually willing to allow the destruction (even by acid rain) of some lakes, if they are not our personal favorites. But the value of the last lake is considerable.

This nonlinearity of values must be included in the scheme by making α in the risk analysis dependent on whether it is the first or last lake we consider. But it is clear that there is far from universal agreement in the country that this should be done. (If there were, there would be no need for this conference and no need for risk analysis, except for a historical account of how the decisions were made.) The risk analysis might tell us how far we can go in reducing acid rain by using cost-effective methods upon which we can all agree, and could then delineate the areas of disagreement for political decisions.

I believe that we can bound the discussion quite well. The proposal has been made that sulfur emissions be cut by a factor of two over the next 10 years, by installing controls on existing power plants. The administration balks, supported by 50 ± 30 percent of the U.S. population because the presently estimated costs are of the order of $20 billion. Yet I know of no one who would not be willing for the U.S. government to pay $200 million in such a cause. This variation is less than the variation in α for the "cost to avert a death" of Table 1 and might be considered encouraging.

PROCEDURES FOR GOOD RISK-BENEFIT ANALYSIS

Having discussed bad risk-benefit calculations, it remains to discuss good ones, or at least to demonstrate that one can exist.

Crouch and Wilson (1982) outline 10 principal steps in a risk-benefit analysis:

1. Defining objectives and goals
2. Identifying assumptions
3. Identifying and measuring risks and costs
4. Identifying and measuring benefits
5. Specifying and highlighting uncertainty
6. Aggregating and comparing risks, costs and benefits
7. Identifying groups at risk and recipients of benefits
8. Addressing inequities in the distribution of risks, costs and benefits

9. Developing decision criteria that include any applicable constraints

10. Communicating and summarizing results

Note that in this set of steps we do not specify a precise mathematical equation to perform Step 6, comparing risks, costs and benefits. Yet the Net Present Value equation is useful. Crouch and Wilson suggest a family of such equations for each local region to address distributional questions.

If, as in this case, α is uncertain, we can recast the question about risks. Can one adjust costs, benefits and risks to make NPV > 0 for a wider variety of values for α? Can one also make NPV > 0 for a large number of individual regions?

This brings up the question of regional equity. Not only is it necessary for an equitable decision for NPV to be greater than zero for all the U.S., it must be greater than zero for any important group. For acid rain the balance can be different for an Ohio resident than for a Vermont one. Ohio produces the pollution which falls on Vermont to cause acid rain. Even now there are two positions. Do we consider the effect of proposed changes in air pollution laws in the absolute, or as changes from the present situation? The present situation is that Vermont is suffering from acid pollution and is having no recompense. If we consider the absolute, then Vermont should recover damages and the polluters, including Ohio, should pay to reduce the pollution. If we argue only about changes from the existing situation, as Ohio senators do, then anyone forced to curtail sulfur emissions would demand recompense. It is not for the risk assessor to decide this question, but he should mention it.

The reader will note that the emphasis here is that the risk-analysis problem must be broadly posed. If posed too narrowly the decision might be more obvious, but wrong.

In Figure 2, I show a general flow chart of risk analysis suggested by Crouch and Wilson (1981). In walking you through the procedure, I note the separate boxes for scientific and economic data. Hopefully these are collected in a reliable and truthful manner independently of the decision and the pressures on the decisionmaker. Not so obvious is the separate box for the risk assessment. This is different from collecting data. It includes making estimates, often with huge uncertainties, where no good data exists. Assumptions have to be made, and in our view should be highlighted.

Making comparisons, such as I often do, is still a non-conclusory role, and can be done by agency staff. However, it is the decisionmaker who must consider all the pressures from various interested parties. The chart can be entered from any point. The decisionmaker can ask for risk assessment. However feedback from different parties concerning the assessment is necessary; otherwise alternative interpretations and possibilities for decisions can be ignored. Moreover I note that, although each box is separate, every player in the scheme must be aware of and understand the others. A decisionmaker should not only understand the numbers but also the uncertainties provided him by his risk assessor.

In some discussions of schemes such as this, the separation of the boxes is over-emphasized. I argue that the person collecting the data must be honest about his

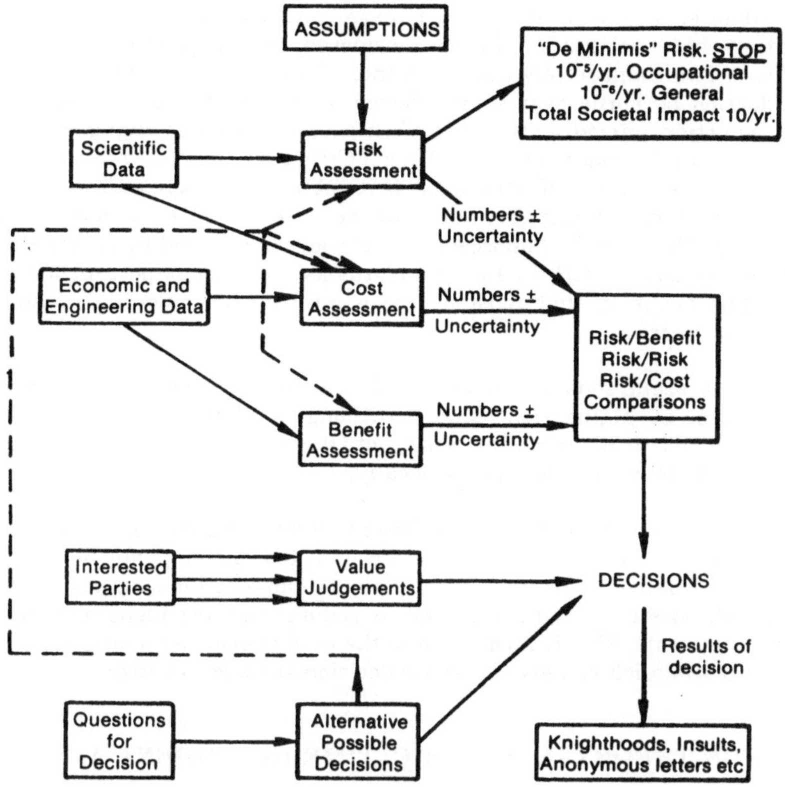

Figure 2. Idealized Scheme for Risk Analysis

science and follow the data no matter where it seems to lead. This is generally agreed. Also desirable is to make the risk assessment, the assumptions for data analysis and the judgments about the data, independently of the decision. This is not easy. Even risk assessors are human and they often have views on what is the correct decision, and their views are probably better than most. But the tendency to bias the assessment to influence the decision must be resisted.

However, having said this, it must be recognized that just as decisionmakers (including some leaders of environmental groups) distrust risk assessors because they are sloppy and leave out some risks, so risk assessors will be angry if their work is misused. There is a story, possibly apochryphal, that in England in Cromwell's time, it was illegal to kiss in public. The draconian penalty was death. But the accused had to be convicted. The juries, a panel of 11 risk assessors, realizing the draconian consequences of their assessment of the truth or falsity of the charge, went beyond their function and denied the evidence of the senses. Who is to say they were wrong? It must also be recognized that there are, no doubt, some actors in this drama who feel they presently have adequate political influence to ensure a decision if no quantitative analysis is provided. Why should they approve risk analysis when the logic might render a decision contrary to their prejudices? As a professional I deplore the approach of such people but have no doubt that it exists.

I will enter into the field of risk manager and leave that of risk assessor, and ask what sensible actions might be considered in the presence of all of this uncertainty. Two immediately come to mind. It would be better to let the costs of sulfur reduction be borne over a long period of time rather than insist on sulfur emission reduction immediately. It is often cheaper to make changes slowly than demand that they be made quickly. Then $\int C(t)\,dt$ will be reduced and NPV will be > 0 for a greater variety of situations. Is it not better to have a slow improvement over two years than an argument for two years followed by a possibility of no improvement? This practical approach was already recognized by Congress when it enacted the new source performance standards as the most important air pollution measure, thus ensuring that most pollution reductions would be gradual as old equipment was retired.

It might seem sensible to build all coal-fired power plants downwind of sensitive areas, in Maine, not Ohio. Although technically sensible, it seems politically difficult. It is Maine residents, not Ohio residents, that want to curb acid rain, and they do not want power plants locally.

Finally, I note our entry on the bottom right hand corner of the figure. A decisionmaker does not necessarily make decisions to improve public health or for the good of the environment. He makes decisions to enhance his own position, gain an honor or knighthood if you will, or perhaps to reduce the consequences to him of error (anonymous letters). This is recognized in the field of decision analysis for business decisions as distinguishing between a good decision and a good outcome.

DECISION THEORY--A DYNAMIC, TIME DEPENDENT APPROACH

I have so far described a static process which is valid at one instant in time. This is posing the risk-benefit decision more narrowly than desirable, an error I condemned earlier. Information can vary. We have already said that the costs of pollution control, C(t), can be reduced if immediate action is not required. The estimate of risk R(t) can be improved with further research. Then the whole balance in the Net Present Value equation can shift. This leads to the suggestion that action be deferred to gain time. Whether this is good or bad depends upon what one does with the time. This brings us to the set of questions:

What is the value of research?

How can one encourage (force) new technology?

This is the realm of decision analysis of which my risk analysis is a static subset. I leave it up to the next speaker, Dr. Warner North, to address this question (North and Balson, 1984).

ACKNOWLEDGEMENTS

My ideas on this subject are not original with myself. While acknowledging that the errors are mine only, I want to thank my colleagues, Drs. Calome, Crouch, Klema, D.G. Wilson and Zeise for their various thoughts and inspirations.

REFERENCES

Carrol, Lewis, 1870, *Through the Looking Glass*, MacMillan: London.

Cohen, B.G., 1980, "Society's Valuation of Life Saving in Radiation Protection and Other Contexts," *Health Physics*, 38, 33.

Crouch, E.A.C. and Wilson, Richard, 1981, "The Regulation of Carcinogens," *Risk Analysis*, 1, 47.

Crouch, E.A.C. and Wilson, Richard, 1982, *Risk Benefit Analysis*, Cambridge, MA: Ballinger Publishing Co.

Crouch, E.A.C., Wilson, R. and Zeise, L., 1983, "The Risks of Drinking Water," *Water Resources Research*, 19, 1359.

Evans, John, Ozkaynak, Haluk, and Wilson, Richard, 1982, "The Use of Models in Public Health Risk Analysis," *J. of Energy and Environment*, 1, 1.

Evans, J.S., Tosteson, T., and Kinney, P.L., 1984, "Cross Sectional Mortality Studies and Air Pollution Risk Assessment," *Env. Int.*, 10, 55.

Lave, L. and Seskin, E., 1970, "Air Pollution and Human Health," *Science*, 69 723.

Morgan, M.G., Morris, S.C., Henrion, M., Amoral, D., and Rish, W.R., 1984, "Technical Uncertainties in Quantitative Policy Analyses--A Sulfur-Air Pollution Example," *Risk Analysis*, Vol. 4, p. 201.

North, D. Warner and Balson, William E., 1984, "Risk Assessment and Acid Rain Policy: A Decision Framework that Includes Uncertainty," paper presented at the Conference on Acid Rain: Economic Assessment, Washington, DC, Dec. 6.

Wilson, Richard, Colome, S.D., Spengler, J.D., and Wilson, D.G., 1980, *Health Effects of Fossil Fuel Burning: Assessment and Mitigation*, Ballinger Publishing Company, Cambridge, MA.

Wilson, Richard, Crouch, E.A.C., Klema, E., 1984, "Comments on the Proposed Air Quality Standards for Particulate Matter," 49FR10408; *Comments on the Impact Analysis*, (internal report, EEPC, Harvard University).

RISK ASSESSMENT AND ACID RAIN POLICY: A DECISION FRAMEWORK THAT INCLUDES UNCERTAINTY

D. Warner North and William E. Balson*

Decision Focus Incorporated
4984 El Camino Real, Suite 200
Los Altos, CA 94022

INTRODUCTION

Uncertainty is the central issue that complicates both qualitative discussion and quantitative economic analysis of acid rain policy.** The available scientific information does not provide a basis for predicting what the consequences of alternative policies will be. Particularly in discussing the impacts of acid rain, scientists and proponents of policy alternatives differ greatly in statements that should reflect the same body of scientific evidence.

Impacts on aquatic ecosystems provide an illustration. The four quotations below are reasonably representative of the divergent views that have appeared in both the scientific and popular literature over the last several years. The first two are taken from Op-Ed columns in a recent Sunday *New York Times*, and the last two from lengthy reports by government agencies.

> Today, hundreds of lakes in New England, the Upper Middle West, the Mountain states, and the Southeast that once teemed with fish have been spoiled by acidification.
>
> (Waxman, 1984)

> The findings [from research] have been far from conclusive. Sulfur dioxide emissions in the Middle West have not been proven to contribute significantly to lake and stream acidity.
>
> (Bagge, 1984)

*D. Warner North, Principal; and William E. Balson, Senior Associate.

This article includes material from D. Warner North, William E. Balson, and Dean W. Boyd, "Acid Deposition: A Decision Framework that Includes Uncertainty" in D. Adams and W. Page, *Acid Deposition-- Environmental and Economic Impacts* (Plenum, to appear 1985).

**In this paper we shall use the term "acid rain" to include all forms of acid precipitation and dry deposition of acidic materials. The term "acid deposition" would be more accurate, but "acid rain" is widely used, and, in particular, is used in the title of this conference and in the name of the sponsoring organization.

> We estimate that 3,000 lakes and 23,000 miles of streams--or about 20 percent of those in sensitive areas--are now extremely vulnerable to further acid deposition or have already become acidic.
>
> (Office of Technology Assessment, 1984)

> In the United States, only in the Adirondack region have adverse effects of acidification on fish populations been observed.... Loss of fish populations has been documented for about 180 Adirondack lakes (out of a total of approximately 2877), although historic records are not available at this time to relate each loss specifically to acidification or acid deposition.
>
> (Environmental Protection Agency, 1984)

In attempting to sort out the confusion concerning impacts on aquatic ecosystems, it is useful to separate two questions: what is the extent of existing damage to surface waters, and what additional damage will occur if additional controls are not placed on sources of sulfur oxide emissions? Neither question can be answered with certainty. There is clear evidence that some lakes (such as those in the LaCloche Mountains in Ontario) have lost their fish populations because of acid deposition from the large smelters nearby at Sudbury, but there is not yet evidence that such effects have occurred on a large scale in the United States. Loss of fish populations has been observed in about 4 percent of the lake acreage in the Adirondack region of New York, and acid deposition is suspected of being a cause. Changes in lake water chemistry have been reported over a much wider area of eastern North America, and there is a potential for much greater damage to freshwater fisheries in the U.S. and Canada than the effects that have been reported to date (OTA, 1984; EPA, 1984).

The same difficulty appears in the assessment of damage to terrestrial ecosystems such as forests. Damage to red spruce has been observed in a few high-altitude locations in the Appalachians. No causal relationship to acid deposition has been established, although many scientists suspect that some aspect of air pollution may be responsible for damaging the trees (OTA, 1984; EPA, 1984; Tomlinson, 1983; Johnson and Siccama, 1983). There is a potential both in the U.S. and in Europe that extensive additional damage to forests could occur. The mechanisms for the damage, the extent of the areas affected, and the time scale for damage to occur are not known, but an intensive research effort is now underway.

Other categories of acid rain impact may be described similarly. There is a potential for damage, but little clear evidence at present that such damage has occurred or will occur if sulfur oxide emissions are not reduced below the levels currently allowed. The present situation in the United States might be described as a polarized debate between proponents and opponents of additional emission controls. Both sides frequently invoke reference to costs and benefits in support of their positions, but assessment of the benefits of sulfur oxide control has not been included in the many reports on acid rain that have been issued by U.S. and Canadian government agencies, the National Academy of Sciences, and various other groups to assist the policy decision process.

ATTEMPTS TO DEAL WITH UNCERTAINTY

For economic assessment methodology to succeed in addressing the extent of the benefits from emissions control, it will be necessary to deal with the large uncertainties in our knowledge of the impacts of acid rain. How can this be done? The traditional approach of the quantitative analyst is to use available data and models to make an estimate. Statistical techniques and sensitivity analysis can then be used to give a range of uncertainty around this estimate in the sense of a confidence limit, or an upper and a lower bound. For acid rain this approach runs into great difficulties. Good data are generally lacking, and modeling of the complex scientific relationships involved in acid rain has not proceeded to the point where consensus exists that the models can be used to predict the relationship between emissions and acid deposition and between acid deposition and consequent impacts.

An approach that has been used in another controversial area of environmental policy is to develop a plausible upper-bound estimate of impacts. For assessing the potential health impacts of chemicals that are suspected of inducing cancer in man, the U.S. Environmental Protection Agency has evolved a set of procedures involving the use of a linearized model for estimating the incremental increase in cancer that will result from lifetime exposure to a unit amount of the chemical in air or water (Anderson et al., 1983). While the EPA procedure remains controversial, the basic ideas are being increasingly accepted across the federal government as a means of assessing the risk that chemicals may pose to human health (Office of Science and Technology Policy, 1985). The method is particularly useful when the plausible upper-bound estimates of risk are relatively low for many of the chemicals of concern; regulatory attention can then be focused on the few chemicals whose risk estimates are high.

A plausible upper bound, or worst-case projection, may not be helpful when there is a potential for large impacts, but a high likelihood that the large impacts will not occur. The extent of potential damage from acid rain cited in OTA, 1984, is high. The likelihood of this damage is not assessed in OTA, 1984.

How likely is the widespread destruction of fisheries and forests from acid rain? Do we have the time to carry out research to determine if such damage is going to occur before committing ourselves to expensive emission controls? Is there a high probability that, if controls are not undertaken now, serious and largely irreversible damage to forest and aquatic ecosystems will result? These questions appear to be central issues of acid rain policy, and they cannot be addressed using traditional quantitative techniques or a worst-case analysis approach. They can be addressed using an approach that deals explicitly with uncertainty.

THE DECISION FRAMEWORK: CONCEPTS AND PRINCIPLES

This paper presents an overview of a framework for policy decisions on acid rain. The framework is intended to aid those responsible for policy in evaluating proposals for placing additional controls on sulfur oxide and other anthropogenic emissions believed to contribute to acid rain, and for mitigating the impacts of acid rain on sensitive ecosystems and materials. The framework is also intended to aid in the evaluation of research programs for organizations such as the Electric Power Research Institute and agencies of the U.S. and Canadian governments, who are spending substantial funds to develop better information as a basis for future policy decisions.

The framework is based on decision analysis, which provides a formal theory for choosing among alternatives whose consequences are uncertain. Decision analysis has been widely taught and practiced in the business community for several decades (Raiffa, 1968; Brown, Kahr, and Peterson, 1974; Holloway, 1979), and it has had some applications to major energy and environmental policy decisions (Howard et al., 1972; NAS, 1975; Manne et al., 1979). It provides a natural way to extend cost-benefit analysis to include uncertainty.

The key idea in decision analysis is the use of judgmental probability as a way to quantify uncertainty. By judgmental probability, we mean the use of probability to summarize judgment on the occurrence of an event or the accuracy of a scientific hypothesis. People commonly refer to the probability (or equivalently, the odds) of rain tomorrow, of winning an election or a sporting event, or of success in a business venture. In these examples, the probability numbers serve to summarize judgments on a multitude of complex factors. The judgments of weather forecasters, sports or political experts, business consultants, or scientists may be good, or they may be poor. What probability provides is a way to describe the likelihood of uncertain events quantitatively, permitting explicit judgments about uncertainty to be incorporated into the decisionmaking process. Where there are disagreements in the judgments, the implications for decisions of these differences can be examined. There is an extensive literature on the assessment of judgment about uncertainty in the form of probabilities (Spetzler and Stael von Holstein, 1975; Kahneman et al., 1982; Wallsten and Budescu, 1983). EPA is now using such methods on an experimental basis in assessing the health effects of air pollutants (Whitfield and Wallsten, 1984).

In developing the decision framework we have used two design principles. The first principle is that the framework should separate risk assessment from the risk management decision criteria. Separation of risk assessment from risk management has been advocated in a recent National Academy of Sciences report on environmental health risks, and this theme has been forcefully articulated by the Administrator of EPA (NAS, 1983; Ruckelshaus, 1983; Ruckelshaus, 1984; EPA, 1984). The second principle is that the decision framework should be capable of examining the specific policy alternatives under consideration by the policy community. This principle is to assure that the framework will be practical and relevant.

The purpose of risk assessment is to summarize available scientific information, including characterization of the uncertainties (NAS, 1983). This is done in the decision framework by including a set of scenarios with associated probabilities that span the range of scientific judgment on the relationship between emissions and acid deposition, and between deposition and impacts such as damage to aquatic and terrestrial ecosystems. The scientific basis for the scenarios and their probabilities of occurrence can be documented and peer-reviewed with the scientific community to assure that they do represent a reasonable summary of available information, including not only the data but also the judgment of experts within the scientific community. Where experts disagree, the extent of their disagreement is made explicit.

Risk management involves making tradeoffs among conflicting economic and environmental objectives in the process of selecting the most appropriate policy alternative. Economic analysis can provide a basis for risk management decisions. If environmental as well as economic objectives are valued in explicit quantitative terms such as monetary costs and benefits, economic analysis can assist decision-makers in consistently identifying the best policy alternatives. In the decision framework, a range of scenarios with associated probabilities will describe the potential consequences of an acid rain policy alternative. Each of these scenarios might be valued in terms of control costs and benefits of damage avoided. Comparisons of the probability distributions over costs and benefits for each of the policy alternatives may help to identify which policy option is preferred.*

The decision framework is then a means of comparing decision alternatives by describing their consequences in two dimensions: likelihood of occurrence, and value, in the sense of costs versus benefits of damage avoided. It should not be thought of as a formula for identifying the best policy alternative, but rather as a vehicle for a process. The judgment of scientists is summarized in the form of probabilities on scenarios describing what will happen under various policy alternatives. This portion of the process is a risk assessment. The risk assessment is then combined with risk management decision criteria to provide a means of evaluating specific control, mitigation, or research policies. The decision framework provides an explicit means of integrating complex scientific information with economic and political judgments in the risk management decision process. The implications of differing judgments can be investigated to ascertain which judgments are most critical to the choice of policy decision. The decision framework facilitates an open style of decisionmaking, in which the interaction of scientific and policy aspects can be made clear to the citizens and groups who have a vital interest in the basis for policy decisions.

DECISION FRAMEWORK: OVERVIEW OF STRUCTURE

We now examine how the decision framework is structured to deal with the complexities of the acid rain policy problem. To understand the effects of alternative control policies, it is necessary to understand the relation that various levels of emission reduction may have on the impacts of acid deposition. The potential changes in impacts must then be balanced with the cost of achieving emission reductions. The comparison of various emission control alternatives is made difficult by several factors:

- There is a large degree of uncertainty about the relationship between emissions and impacts.
- It is difficult to compare the value of changes in impacts to the costs of emission reductions.

*In this paper we use the decision criterion of maximizing expected net benefit to identify the preferred decision alternative. Decision analysis provides a more general decision criterion of maximizing expected utility, which can incorporate aversion to risk. In our opinion, the key issues of acid rain policy are the probability of severe acid rain-induced damage scenarios and the evaluation of the benefit of avoiding such damage compared to the cost of emission controls. For conceptual and computational simplicity we use the expected value criterion; the analysis is easily extended to include risk aversion using techniques described in Raiffa, 1968; Brown, Kahr, and Peterson, 1974; Holloway, 1979.

- People involved in assessing control and mitigation policies have different opinions about the evaluation of costs and impacts and may differ in their judgments of the uncertainty in costs and impacts.

- The uncertainty in the relationship between emissions and impacts will be resolved only over a lengthy period of time.

Three stages can be distinguished in the relationship between control alternatives and impacts, as shown in Figure 1. First, there is the effect that control strategies will have on emissions. Then changes in emissions must be related to changes in acid deposition. Finally, changes in acid deposition must be related to changes in the various impacts that can be identified, such as decreased forest productivity and the loss of fish in lakes and streams. There is scientific uncertainty about each of these stages. Relatively little is known about how specific changes in acid deposition will affect changes in impacts. The estimates given by respected scientists vary over a wide range. There is somewhat less uncertainty regarding how changes in emissions will affect changes in deposition; however, the range of uncertainty is still quite large, due primarily to the complex nature of the chemical transformations that occur in the atmosphere. There is comparatively little uncertainty about how control strategies would affect changes in emissions. Accordingly, in implementing the framework the importance of uncertainty in the other two stages, emissions to deposition and deposition to impacts, has been stressed.

At present, the scientific evidence regarding the effects of emissions is contradictory and subject to different interpretations by various experts. The decision framework allows an investigation of the implications of the differing assessments and evaluates the importance of the disagreements in terms of their effects on the choice of a control or mitigation strategy. Many experts who disagree about the interpretation of the current state of knowledge, agree that in five to ten years many of those disagreements could be settled. Thus, in the decision framework, the choice is characterized as one in which we may act now, at a large cost, and accept the possibility that emission reduction will have little beneficial impact. Alternatively, we can wait five to ten years to act on better information that may be available, and accept the possibility that damages may occur during that period. In each case, there is a possibility that the decision will turn out to have been incorrect. From our current state of knowledge, we cannot be sure.

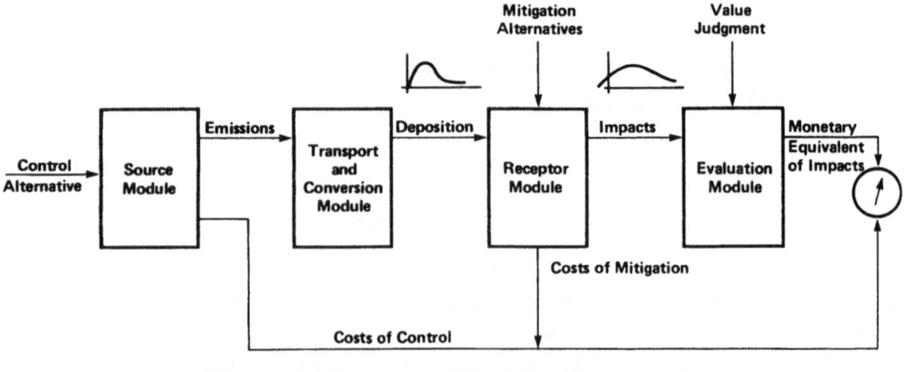

Figure 1. Overview of Decision Framework.

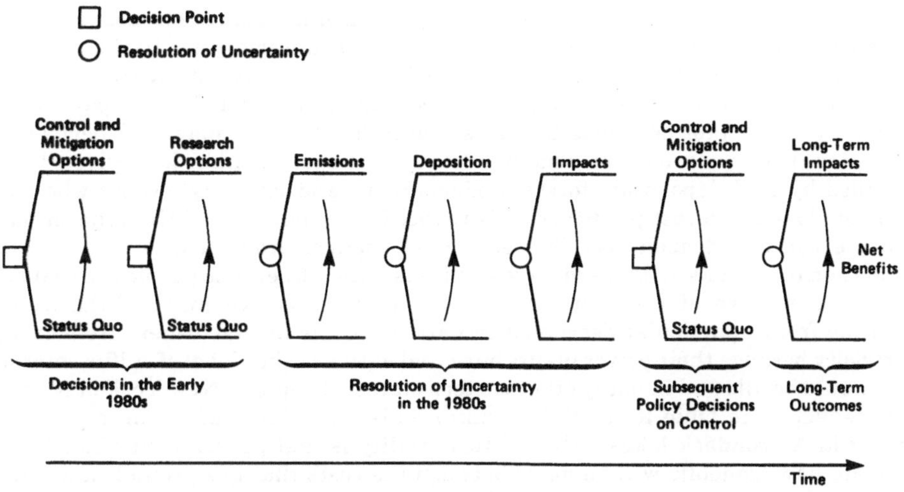

Figure 2. Decision Tree for Acid Rain Policy.

The strategies that are available and the resolution of uncertainty at different points in time are represented as a decision tree (Figure 2). A decision tree is simply an efficient way of describing a set of scenarios. Each particular set of decisions and outcomes representing how uncertainty could resolve comprises a scenario. Each scenario answers a "what if?" question corresponding to "what if a particular policy were chosen," followed by a particular change in deposition, and finally by a particular change in impacts.

The decision tree of Figure 2 is a generic representation of the time sequence of choices among decision alternatives and the resolution of uncertainty in the areas enumerated in the framework of Figure 1. The first two stages, shown at the far left of the figure, are the decisions within the next few years on control and mitigation options and on a national research program on acid deposition. The next two stages represent resolution of uncertainty on the relation of acid deposition to emissions and the relation of impacts to acid deposition as the research program is carried out and new scientific knowledge is obtained. Next comes a decision point in the late 1980s or early 1990s when national policy on control and mitigation would be reassessed and an alternative chosen on the basis of the new information that has recently been made available. Further resolution of uncertainty on deposition and on impacts of acid deposition then follows.

The decision tree of Figure 2 provides a rich sequence of scenarios describing the decisions and outcomes characterizing national policy on acid rain. It includes two stages of decisionmaking, one with present information and one with the information that might become available five to ten years hence following an extensive research program. The decision tree explicitly includes the option of taking action now to control emissions or mitigate the effects of acid deposition and the option to wait until better information becomes available in five to ten years. The effect of today's research funding decisions and the choice of emphasis in the research program may strongly affect what information becomes available in the next five to ten years, and this interaction is explicitly considered in the decision tree framework.

The decision tree approach provides a useful separation between value judgments on costs and benefits and judgment about uncertainties in the impacts of acid deposition. Each scenario in the decision tree may be considered as having impacts on a number of concerned parties: consumers who may have to pay more for electricity because of decisions to impose controls on power plants, fishermen and recreational property owners who stand to lose if sport fishing in a given lake is degraded by acid deposition, forest products firms and property owners who suffer economic losses if forest productivity is reduced, and members of the general public who are concerned about possible ecological changes from acid deposition. The evaluation of impacts on these diverse parties is difficult because people see that some parties bear more of the costs while other parties receive more of the benefits resulting from a particular decision alternative. People in Ohio benefit from cheaper electricity because their power plants burn coal with a higher level of sulfur emissions than would be allowed in many other eastern states. People in New York may benefit from reduction in Ohio River Valley sulfur emissions if the reduction improves the fishing in Adirondack lakes. The political reality is that government officials must evaluate how tradeoffs will be made between the costs that one group bears and the benefits that another group receives. Issues of equity and property rights make such value judgments extremely difficult. It is useful to separate these value judgments from the uncertainty in the effects that long-range transport of sulfur and other pollutants may cause. The decision framework accomplishes this desired separation between the answer to the question of what will happen under a given choice of control and mitigation strategies and the societal evaluation of what each outcome is worth.

A DECISION TREE CARICATURE

A decision tree of the complexity of Figure 2 can include hundreds of thousands of paths or scenarios defined by different combinations of decisions and outcomes at each stage. Carrying out calculations on decision trees of this size usually involves the use of digital computers, and our analysis has in fact been implemented in this fashion, as a FORTRAN code called ADEPT (Acid DEPosition decision Tree program, which is available from EPRI) (Balson, Boyd, and North, 1982). We can illustrate the form of these calculations with the highly simplified decision tree shown in Figure 3, which has only six paths or scenarios. This decision tree can be regarded as a caricature of the complex tree shown in Figure 2. Almost all of the structure of the decision as described previously has been eliminated, leaving only two choices: (1) action to further reduce emissions of sulfur oxides from current levels and (2) no additional control on sulfur oxides or other emissions. Uncertainty is represented on the extent to which the emission reduction lowers acid deposition and on the relationship between the acid deposition and the extent of long-term ecological impacts, which could be minor or large.

To some extent, this simplified decision tree illustrates the current polarized debate about acid deposition. Proponents of control legislation have argued that extensive reduction in sulfur oxide emissions can only be accomplished if action is taken now, and that acid deposition resulting from continued emissions may cause far more extensive damage to lakes, streams, and forests than the effects observed to date. Their adversaries have argued that the links between sulfur emissions, acid deposition, and ecological damage have not been established clearly, and that emission reduction would incur large costs and might achieve little benefit in avoiding damage from the impacts of acid deposition.

We have represented these arguments in the form of six scenarios, four proceeding from an additional control alternative and two from an alternative of no additional control. For the additional control alternative, let us assume that a 50-percent reduction in sulfur emissions from coal-fired power plants in the Ohio River Valley can be achieved with fuel switching and flue gas desulfurization at a cost of $2 billion per year (in constant dollars). Two outcomes are assumed for the reduction in acid deposition in sensitive downwind receptor areas such as the Adirondacks if sulfur emissions are reduced 50 percent by Ohio River Valley power plants. Averaged over the year, acid deposition in the Adirondacks corresponds to about 60 percent sulfate, 30 percent nitrate, and 10 percent other species. If we assume that virtually all of the sulfate in the Adirondacks comes from power plants burning coal in the Ohio River Valley and that the relationship between emissions and deposition is linear, then a reduction in the sulfur emissions from these sources would lead to a reduction in acid deposition of about 30 percent. However, if local sources of sulfur oxides play an important role or if the emissions-deposition relationship is nonlinear, then the reduction in emissions from Ohio River Valley power plants could result in a much smaller change in the level of acid deposition in the Adirondacks, for example, ten percent. For illustration, we assume only these two possibilities, and each is judged to have a probability of 50 percent of being correct.

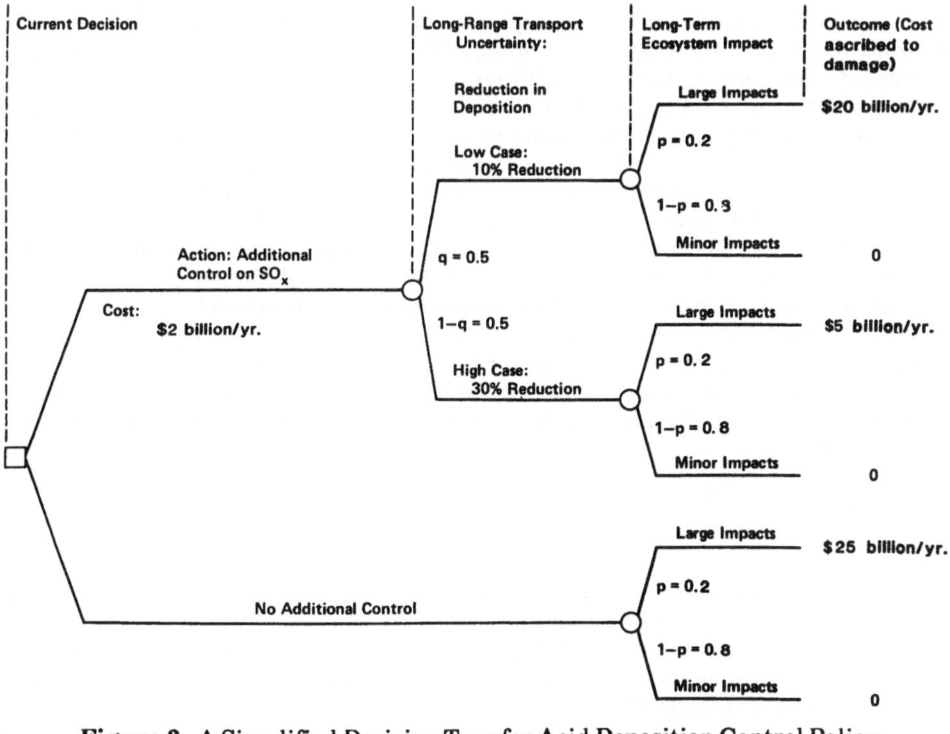

Figure 3. A Simplified Decision Tree for Acid Deposition Control Policy.

We likewise assume only two outcomes for ecosystems impacts: large impacts and minor impacts. The "minor impacts" outcome implies that the damage to surface waters, soils, forests, and material property will be no worse than the effects that have been observed to date. The "large impacts" outcome assumes that continued high levels of acid deposition will lead to damage that is far worse, causing loss of sport fisheries, reduced productivity of forests, and other adverse impacts across much of the northeastern United States and eastern Canada. We assume for our illustrative calculation a probability of 20 percent that this large impact scenario is going to occur.

We shall also assume that we can describe each of the ecological system outcomes in terms of an equivalent monetary value that decisionmakers agree the United States would be willing to pay to avoid such an outcome. For simplicity and convenience, we take the monetary equivalent for the three minor damage scenarios to be $0. For the large impacts, we assume a level of $25 billion per year (in constant dollars) for the case of no reduction in deposition. In the ADEPT model, such values are calculated based on estimates over time of the extent of forest and surface water acreage damaged by acid deposition. If deposition is reduced by 30 percent, we assume that the extent of the damage is reduced more than proportionally, so that the equivalent monetary damage is reduced to $5 billion annually. If deposition is reduced only 10 percent, the assumed monetary equivalent of the impact is reduced from $25 billion to $20 billion.

Once the scenarios have been described in terms of probabilities and values, the decision tree can be used to obtain valuable insights in comparing the decision alternatives. As a means of evaluating the alternatives, we will compare the sum of the control cost plus the expected or probability-weighted average of damages. The expected damage is computed by multiplying the probability of each outcome by the monetary damage if that outcome occurs. Therefore, the expected damage corresponding to the no-additional-control alternative is $5 billion per year, which is 20 percent times $25 billion plus 80 percent times $0. The expected damage for the additional control alternative is computed by multiplying the probability times the monetary value for four cases and summing: (0.5)(0.2)($20) + (0.5)(0.2)($5) + 2(0.5)(0.8)($0) = $2.5 billion per year. A cost of $2 billion per year for control is added for the additional control alternative: $2.5 + $2 = $4.5 billion per year. The least total annual cost is therefore obtained by choosing additional control with an expected total cost of $4.5 billion compared to the $5 billion expected damage with the no-additional-control alternative. We can see from this evaluation that the total costs for the two alternatives are close. Reducing the probability of the large impact scenario from 0.2 to 0.15 would make the no-additional-control alternative have the lower total expected cost.*

*The calculation of the expected value for each alternative is made in the same way as described in the above paragraph, but now using 0.15 instead of 0.20. The expected damage for the no-additional-control alternative is (0.15)($25) + (0.85(0) = $3.75 billion, and the expected damage for the additional control alternative is calculated from the equation above with 0.15 and 0.85 replacing 0.2 and 0.8 in the terms on the left hand side, yielding a total of $3.875 billion.

An important concept in decision analysis for evaluating research and other information gathering activities is the expected value of information. Suppose we could resolve which impact case and which long-range transport case were true before choosing between control and no additional control. How much would it be worth to gain this information before making the choice? We can answer this question with the probabilities and values in the decision tree of Figure 3 by making another expected value calculation. This calculation corresponds to reversing the order of the stages in the tree; the expected value is computed assuming that for each possible outcome on ecosystem impacts, the decision alternative will be chosen with the lowest total of control costs plus damage. If large impacts will occur, the least costly choice is additional control. If minor impacts will occur, the least costly choice is no additional control. The probabilities were given above: 0.2 for large impacts and 0.8 for minor impacts. Recall that in the absence of information, the least costly decision was additional control. If the information indicates large impacts, additional control will remain the least costly decision alternative. But if the information indicates minor impacts will occur, no additional control will be the least costly alternative, and a savings of \$2 billion in control cost can be achieved. The expected (i.e., probability weighted average) savings from making the decision after the new information is available is then $0.8 \times \$2.0 = \$1.6$ billion per year. Figure 4 shows the expected value of perfect information (the expected savings if the decision could be made knowing which impact case is correct) as a function of the probability p of large ecosystem impacts. We see in Figure 4 that the maximum for the expected value of perfect information occurs at $p = 0.16$, the point at which control and no additional control yield equal total expected costs.

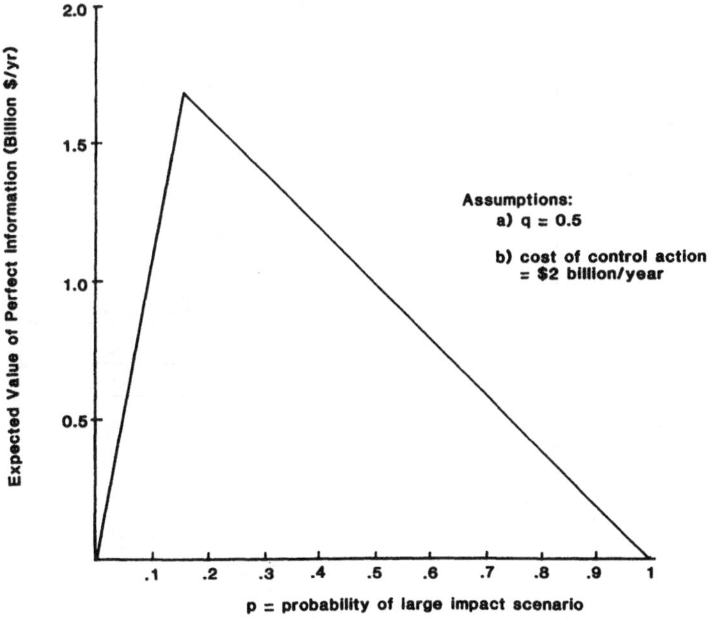

Figure 4. The Expected Value of Perfect Information.

Expanding the Caricature: Act Versus Wait

The decision we have just examined was a choice between action, additional control on sulfur oxide emissions, and no additional control. A more accurate characterization of the decision facing policymakers is a choice between acting now and waiting. We show this choice in an expanded caricature decision tree in Figure 5. The no-additional-control alternative has been replaced by a wait alternative, where we have assumed that perfect information would be available after ten years and that additional control (50-percent reduction in sulfur emissions) would be implemented in the case of large impacts. We observe that the value judgments characterizing scenarios now become more complex, because we must consider the incremental damage that occurs by waiting ten years before implementing additional control of emissions as opposed to acting now. The damage outcome for the scenario in which additional control is imposed after ten years will be less than the damage under no additional control, and more than the damage if additional emissions control is implemented immediately. For the low (10 percent) reduction in deposition case, let x_1 be the increase in damage from delayed control compared to immediate additional control. Since the damage cannot exceed that for the no-additional-control scenario, $x_1 < \$5$ billion. Similarly, let x_2 be the increase in damage from delayed control compared to immediate additional control in the high (30 percent) reduction in

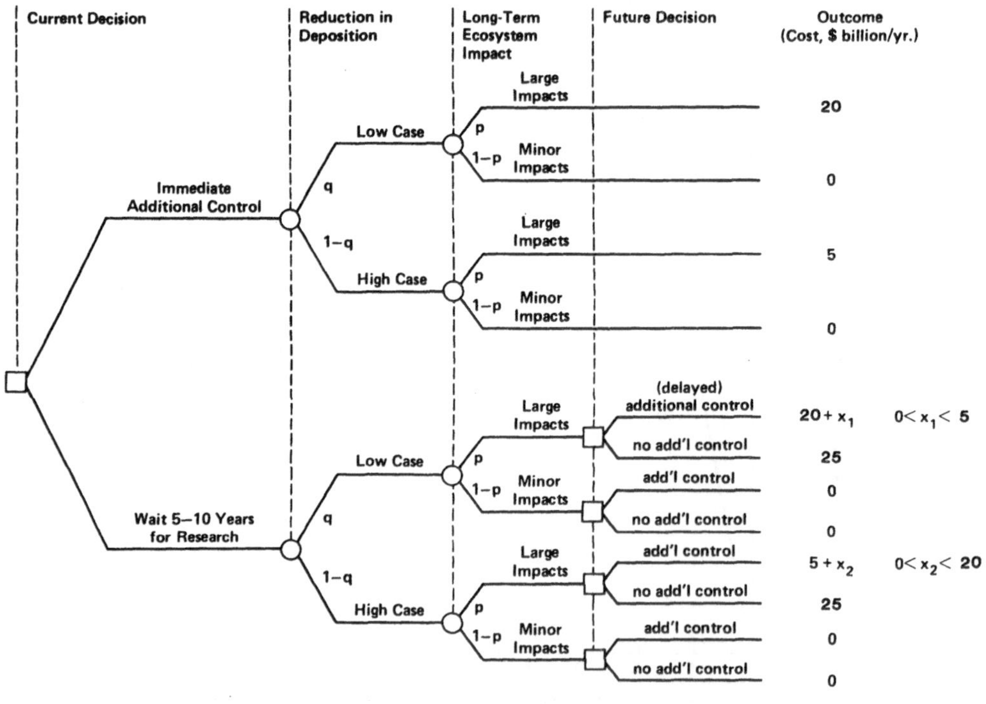

Figure 5. Acid Deposition Decision Tree: Expanded Caricature

deposition case, and we see that $x_2 < \$20$ billion. Once assessments of x_1 and x_2 are made, the analysis of the tree can be carried out in the same manner as before.

Interpreting the Caricature

The caricature decision tree portrays the reason for taking action as the threat of uncertain but potentially very serious adverse ecological consequences that may occur if high levels of acid deposition continue into the future. The damages that have occurred to date may be most important as an indicator, similar to the canary whose death warns miners that the air in the mine can no longer support them.

The choice of whether to take action now to reduce emissions is therefore similar to a decision to buy insurance against a possible future disaster. The large impact outcome is uncertain, and in the analysis we have described its likelihood by a probability that summarizes scientific judgment. Establishing this probability will be difficult because scientists differ in their interpretations of the information now available. However, it may be possible to obtain general agreement that the probability lies within a given range, and within this range it may be clear whether purchasing the insurance is a good idea.

If the decision were between additional control and no additional control, we might expect that the decision would be very sensitive to the probability of the large impact outcome. But the opportunity exists to impose controls at a later time. We have rather arbitrarily taken this time to be ten years, but we can vary the time assumed for the future decision point to see if this assumption makes a difference. The consequence of not taking action now is that the ecological effects under the large impact scenario may be made worse by waiting. However, if we wait, we might learn that the large impact scenario is less likely, or even extremely unlikely. We may alternatively find less costly ways to reduce emissions or mitigate the impacts of acid deposition.

THE DECISION TREES USED IN THE ADEPT MODEL

We now return to the generic decision tree of Figure 2 to determine how we should proceed in constructing an implementation of the decision framework to address the questions we have discussed above. We recognize that there are various assumptions we could make regarding how fast uncertainty with respect to ecological impacts and long-range transport will be resolved. One limiting assumption is that no resolution of uncertainty will take place before the second decision point. The decision tree then has the form shown in Figure 6. A major difference between the second decision point and the first could be the availability of a new control technology, such as the lime injection multistage burner (LIMB), that is more effective or less costly than the technologies now available.

An assumption at the other extreme is that uncertainties on ecological impacts and long-range transport will be essentially resolved within five to ten years, so that the second decision can be assumed to take place under certainty, or with perfect

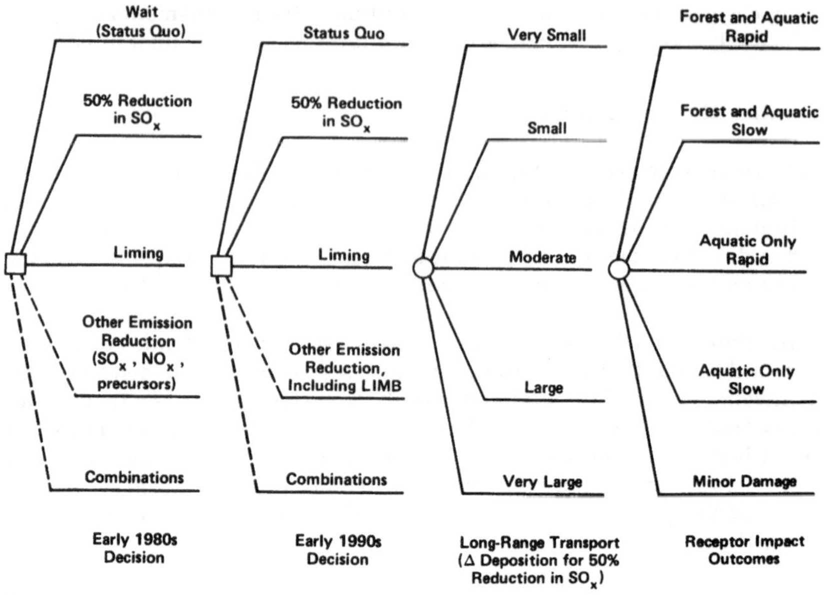

Figure 6. A Decision Tree with No Resolution of Uncertainty by the Early 1990s.

information.* This decision tree compares the ecological consequences of waiting before taking action with the value of perfect but delayed information. Our perception is that there is widespread agreement within the scientific community that much could be learned in the next five to ten years regarding the ecological consequences of acid deposition and the long-range transport relationships between emissions at a source region and deposition in a receptor region. The policy debate as we interpret it is whether we can afford to wait five to ten years before taking action. We have therefore chosen the decision tree shown in Figure 7 as the basic decision tree in ADEPT for examining control and mitigation alternatives. The tree is essentially the same form as the expanded caricature we have described above, except that more decision alternatives have been added. Five rather than two cases have been used to represent long-range transport uncertainty, and five rather than two cases have been used to represent uncertainty on ecological impacts; a total of twenty-five combinations instead of four.

*While complete resolution of uncertainty or perfect information is a limiting case, it corresponds practically to the situation where new research results have reduced the uncertainty sufficiently so that the choice among alternatives is clear. Analysis of the value of perfect information is a simple calculation that can be used to guide the evaluation of real, imperfect information-gathering alternatives (Raiffa, 1968; Brown et al., 1974; Holloway, 1979; Howard et al., 1972; Balson, Boyd, and North, 1982). The more complex decision tree of Figure 8 described below is used for analysis of imperfect information-gathering activities (Balson, Boyd, and North, 1982).

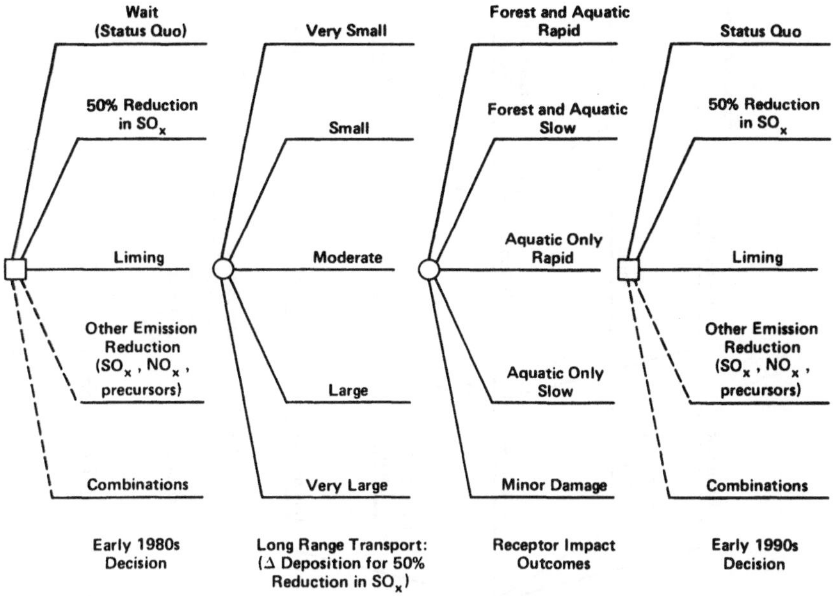

Figure 7. A Decision Tree with Full Resolution of Uncertainty by the Early 1990s.

Figure 8 shows a more complex version of the generic decision tree of Figure 2, which incorporates the decision on alternative research programs and the research results obtained from these programs. In this research emphasis decision tree, we assume that additional information about long-range transport and ecological effects may become available within ten years. Uncertainty is thus resolved in two stages, some in the first time period before the second decision point and some afterward. Large research efforts give a higher expectation of resolving the uncertainty early, but for all the research options the research results achieved may be inconclusive or wrong. The research emphasis decision tree in ADEPT permits the evaluation of research programs that should provide a better basis for future decisions on control and mitigation. The research emphasis decision tree is substantially more complex than the basic tree, but the calculations are still easily made on a computer. With the branches represented by solid lines in Figure 8, there are over six thousand scenarios in this tree. The development of consistent probabilities for the research results and the outcomes on long-range transport and ecological effects requires a considerable amount of thought about the effect of specific research findings on the state of scientific knowledge.

Use of the decision framework relies heavily on the assessment of judgmental probabilities. The ADEPT decision tree models require that such probabilities be provided as input data. The assessment process is a difficult and subtle art, especially as in this application where the judgments concern issues of great complexity and cut across many scientific specialties (Spetzler and Stael von Holstein, 1975; Kahneman

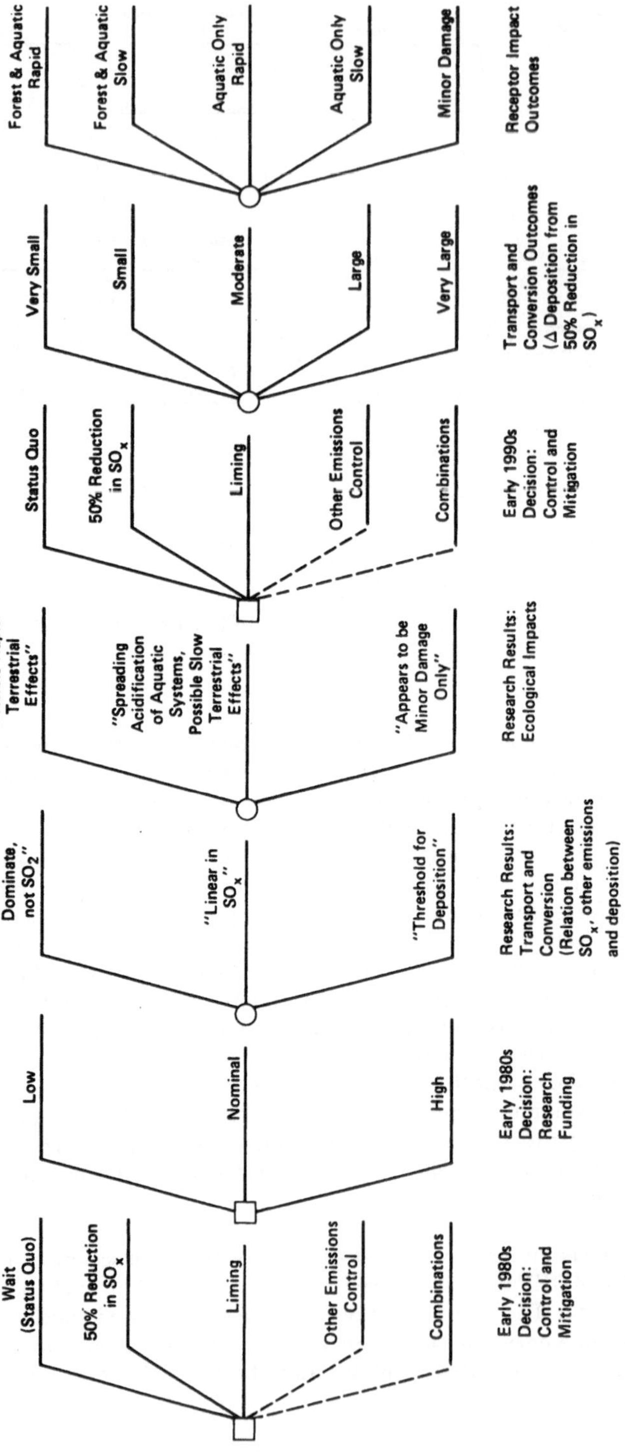

Figure 8. A Decision Tree with Resolution of Uncertainty Dependent on Research Funding Decision.

et al., 1982; Wallsten and Budescu, 1983). The key to success in using judgmental probability is the credibility of the analysis process. The expert whose judgment is being assessed must understand the assessment process and the way in which his or her judgment is being used. This requirement implies that substantial time will be needed for communication between analysts responsible for assessing probabilities and the experts whose judgment is being sought.

The other sensitive aspect for both the basic tree and the research emphasis tree is the characterization of acid deposition impact outcomes. The assumptions and models used in ADEPT are highly modular and easily understood graphically (Balson, Boyd, and North, 1982; Balson and North, 1983). The reduction in emissions resulting from a specific control strategy is phased in over time. Various linear and nonlinear assumptions about the change in deposition that results from a given change in emissions can be utilized. The time pattern of surface water and forest acidification that occurs under the different ecological impact scenarios can be varied. Each of these relationships can be changed within a wide range of possibilities, and each is modular within the ADEPT code so that it can be replaced by an entirely different set of assumptions relatively easily. The monetary equivalents per damaged acre of surface water and forest are important and potentially controversial, but the analysis can easily be repeated over a wide range for these value judgments to show the implications of differing opinions for selecting the alternative with the least expected total cost.

USE OF THE DECISION FRAMEWORK

In the two years since the acid deposition decision framework and the ADEPT computer implementation were developed, the framework has received considerable attention and acceptance. It has been used within the Electric Power Research Institute (EPRI) in planning research program strategy and for illustrative calculations of the costs and benefits to individual states of acid deposition control policies (Balson and North, 1983). The ADEPT model has also been licensed by EPRI to agencies of the U.S. Government, including the Department of Energy and the Environmental Protection Agency. An extensive review of ADEPT has been carried out for the Environmental Protection Agency (Davidson, et al., 1984).

The fundamental conflicts and differences in perception regarding acid deposition will not be resolved easily, and any attempt at consensus building will be regarded with suspicion by many in the scientific and the policy communities. Nonetheless, we believe that the decision framework and the ADEPT model offer great potential for consensus building among the electric utility industry, environmental interest groups, government agencies, and other organizations concerned about acid deposition, and about similar complex and divisive environmental issues.

ACKNOWLEDGEMENT

This paper is based upon research carried out for the Electric Power Research Institute under EPRI Research Project 2156-1.

REFERENCES

Anderson, E.L., and the Carcinogen Assessment Group of the U.S. Environmental Protection Agency, 1983, "Quantitative Approaches in Use to Assess Cancer Risk," *Risk Analysis*, 3 no. 4:277-295.

Bagge, C.E., October 28, 1984, "The Case Has Yet to Be Proven," *The New York Times*.

Balson, W.E., Boyd, D.W., and North, D.W., 1982, *Acid Deposition: Decision Framework, Volume I: Description of Conceptual Framework and Decision Tree Models*, report prepared by Decision Focus Incorporated for the Electric Power Research Institute, EPRI EA-2540, Palo Alto, Ca.

Balson, W.E., and North, D.W., 1983, *Acid Deposition: Decision Framework, Volume 3: State-Level Application*, report prepared by Decision Focus Incorporated for the Electric Power Research Institute, EPRI EA-2540, Palo Alto, Ca.

Brown, R.V., Kahr, A.S., and Peterson, C.R., 1974, *Decision Analysis for the Manager*, Holt, Rinehart, and Winston, New York.

Davidson, C.I., et al., 1984, "A Conceptual Framework for Integrated Assessments of the Acid Deposition Problem," Draft Final Report to the Acid Deposition Assessment Staff, U.S. Environmental Protection Agency and Acid Deposition Program, North Carolina State University by the Center for Energy and Environmental Studies, Carnegie-Mellon University, Pittsburgh, Pa.

Holloway, C., 1979, *Decision Making Under Uncertainty: Models and Choices*, Prentice-Hall, Englewood Cliffs, N.J.

Howard, R.A., et al., 1972, "The Decision to Seed Hurricanes," *Science*, 176:1191-1202.

Johnson, A.H., and Siccama, T.G., 1983, "Acid Deposition and Forest Decline," *Environmental Science and Technology*, 17 no. 7:294A-305A.

Kahneman, D., et al., eds., 1982, *Judgment Under Uncertainty: Heuristics and Biasis*, Cambridge University Press, Cambridge, Ma.

Manne, A., et al., 1979, "Energy Policy Modeling: A Survey," *Operations Research* 27 no. 1:1-36.

National Academy of Sciences, Commission on Natural Resources, 1975, *Air Quality and Stationary Source Emission Control*, U.S. Government Printing Office, Washington, D.C.

National Academy of Sciences, Committee on the Institutional Means for Assessment of Risks to Public Health, 1983, *Risk Assessment in the Federal Government: Managing the Process*, National Academy Press, Washington, D.C.

Office of Science and Technology Policy, 1984, "Chemical Carcinogenics; A Review of the Science and its Associated Principles," *Federal Register*, 50 no. 50:10372-10442.

Office of Technology Assessment, 1984, *Acid Rain and Transported Air Pollutants: Implications for Public Policy*, OTA-0-204, U.S. Congress, Washington, D.C.

Raiffa, H., 1968, *Decision Analysis: Introductory Lectures on Choices Under Uncertainty*, Addison-Wesley, Reading, Ma.

Ruckelshaus, W.D., 1983, "Science, Risk, and Public Policy," *Science*, 221:1026-1028.

Ruckelshaus, W.D., February 18, 1984, "Risk in a Free Society," *Risk Analysis*, 4:157-162.

Spetzler, C.S., and von Holstein, C.A.S., 1975, "Probability Encoding in Decision Analysis," *Management Science*, 22:340-358.

Tomlinson, G.H., 1983, "Air Pollutants and Forest Decline," *Environmental Science and Technology*, 17 no. 6:246A-256A.

U.S. Environmental Protection Agency, 1984, *The Acidic Deposition Phenomenon and Its Effects: Critical Assessment Review Papers*, Volume II:, Effects Sciences, EPA-600/8-83-016bF.

U.S. Environmental Protection Agency, 1984, *Risk Assessment and Management: Framework for Decision Making*.

Waxman, H.A., October 28, 1984, "Why Do We Ignore the Evidence?" *New York Times*.

Wallsten, T.S., and Budescu, D.V., 1983, "Encoding Subjective Probabilities: A Psychological and Psychometric Review," *Management Science*, 29:151-172.

Whitfield, R.G., and Wallsten, T.G., 1984, *Estimating the Risks of Lead-Induced Hemoglobin Decrements under Conditions of Uncertainty: A Methodology, Pilot Judgments, and Example Calculations*, report prepared by Argonne National Laboratory for the Office of Air Quality Planning and Standards, U.S. Environmental Protection Agency.

RESPONSE TO RICHARD WILSON AND D. WARNER NORTH

David Hawkins

Senior Staff Attorney
Natural Resources Defense Council, Inc.
1725 I Street, N.W.
Washington, DC 20006

Both of the presentations have at their core a discussion of cost-benefit analysis. I would like to take a few minutes to react and to explain some of the concerns that the environmental groups in particular, and I think the public generally, have about using this as the model for decisionmaking in the environmental arena. I might point out that there are several models that one can hypothesize in addition to cost-benefit models. There are models which I will label "ecosystem conservative" models, and a third type of model that I will call the "reasonable person-good judgment" model, or if that sounds too normative a label perhaps the "risk averse/cost-conscious" model.

What are some of the problems that the environmental community have with cost-benefit analysis? I would suggest there are two key questions underlying the usefulness of these models in decisionmaking. The first is whether we can accurately account for costs and benefits and in particular whether we can accurately capture all of the ecological risks to the same degree that we can identify abatement costs. There is a tendency underlying cost-benefit analysis to assume that these risks can be captured. I would submit that this belief is more appropriate to the engineering context than it is to the biological systems context.

In particular, I would submit that a rapid capturing of all ecological risks is extremely unlikely in the biological system context, for a variety of reasons. I will just mention one in particular; biological systems have feedback loops. These feedback loops can mediate and tend to disguise effects and stresses on the systems. Human beings do not spot these stresses or do not regard the effects which they do see, as being stresses which result from these feedback loops. How long these feedback loops operate, human beings cannot assess accurately until the processes have ended. This is a serious issue and it suggests that the accumulation of information about ecological risks is a lot slower than the accumulation of knowledge about risks in the engineering context.

Nor is it easy to create a risk inventory that will satisfy those that question the need for controls. Typically, those questioning the need for controls want first to have the effects identified, second to have agreement that they are adverse, third to prove the causal links to certain of man's activities, and fourth to establish dose-response functions. But these are not easy jobs, they do not come quickly, and they do come at a price of a long wait and perhaps a wait that is much longer than appropriate given the size of the threats that may be involved.

Dick Wilson, in his talk, noted how easy it is to understate the benefits in undertaking one of these analyses and how the analyses are not necessarily free of values when calculating benefits. He has pointed out that one caricature of the understated-benefit approach is what you would call the Stockman approach. I would agree with that, and I would agree that the Stockman approach does characterize a lot of the cost-benefit analysis that goes on. The question that I would raise is, "What does it say about the neutrality of the administration's embrace of cost-benefit analysis, when they have made Mr. Stockman the cost-benefit analysis (CBA) cop, more accurately, the Chief of Police?"

Warner North proposes a structure that would compensate for some of our ignorance by essentially asking experts to assign their judgment about the probability of damage. There are problems with this too. There is nothing wrong with an organized system of assembling what we know and what we don't know, but we do need initially to decide on the effects to which we are going to ask the experts to assign probabilities. If we leave out an effect, we are going to wind up understating the risks. The assumption underlying this technique is that we have enough knowledge to be able to identify all the questions that we ought to ask. As I said a moment ago, I do not think that this is necessarily a valid assumption.

Second, there is a lot of controversy about the theory of probability encoding which lies at the heart of the technique. Essentially it involves converting people's opinions into numbers. Doing that allows the probabilities to be averaged, but it is not clear that you necessarily produce a better answer than you would by simply hearing people express their opinions and answer questions about the basis for those opinions. The average that you get depends on the sample you select, and you can have ten people that an objective observer would regard as "kooks" and one person that somebody would regard as knowing what he or she was talking about, and you can produce an average that doesn't necessarily give you a very good assessment of the risk you are analyzing.

Well, beyond risk and cost identification, cost-benefit analysis requires that a value be assigned to these effects. And I would submit that techniques for assigning these values are not neutral. I will just mention a few concerns. For risks without markets, one approach the economists like to use is the willingness-to-pay index, and one of the willingness-to-pay indices that they like to look at is the wage premium that is demanded for hazardous jobs. In my view, these studies really measure the degree of economic distress a person is experiencing and the number of people suffering from that distress. A person with very few job opportunities and little economic security is not going to be able to demand a high wage premium no matter how risky a job is, particularly when there are long lines of people right behind him competing for the same job. So if we believe that our society should use economic distress as a tool for distributing risks, we would be comfortable with this evaluation technique. But if we disagree with that, we won't be comfortable. The point is that it's not a neutral technique.

As to compliance costs, there is an assumption built-in that the size of the compliance costs is independent of the initial choice to use cost-benefit analysis as a decisionmaking approach. But it is not. Consider for a moment a different model, the cost-internalization model. If we have a mandate to internalize the costs a firm imposes on a society, we are going to produce a managerial incentive to invest money, research and development dollars, which represent hard money for which the corporation has many competing needs. We will produce an incentive to invest those dollars in an effort to minimize the expense of internalizing the costs. If internalizing the costs is a given, the company has to do it; it is going to have an incentive to spend money to find cheaper ways to comply.

Cost-benefit analysis produces a different result. It says, "If current assessments of compliance costs are high enough, then we don't require the firm to internalize the costs." In that case, we will not take regulatory action, and thus will remove any incentive for the firms to invest the money that would be necessary to reduce the costs in the first place. I hope you see the circularity problem here.

Now I'd like to turn to a second major concern. That is, even if cost-benefit analysis could accurately capture all the costs and benefits, there is a lot of objection to the underlying premise that the state should legalize activity which is regarded as an expropriation of personal rights, simply because of a conclusion by the state that net societal benefits appear to be maximized. Stated another way, pollution control laws are the political expression of what the public believes is proper and improper behavior. I would submit that the public regards pollution as an activity that harms people and that it is government's job to prevent people from acting in ways that harm others. It is not government's job to rationalize the harm by resorting to cost-benefit analysis. Well, all well and good, but if not cost-benefit analysis, then what?

Let me describe two other models that could be talked about and I won't talk about them very long. One is what I've called an ecosystem conservative approach. This model would basically hold that, if any of man's activities appear to result in a large change in any natural system, we assume it's bad and we prohibit it. As an ideal for the long run, that might be a wise approach. It isn't too useful in the 20th century at least, because it appears to prevent many activities of man that are regarded as valuable. So it doesn't really provide a very useful working tool for making decisions in the real world today, even if it may turn out five centuries from now to have been a wise approach.

Let us consider the second approach, which I have called the risk-averse, cost-conscious approach. In this model one starts by examining the magnitude of man's activities, and in that sense it is similar to the ecosystem-conservative approach. For example, we would note in the acid deposition context, that man is emitting sulfur into the air in the eastern United States at a rate which people estimate to be anywhere from 20 to 100 times the natural burden. That's a big change and it ought to send up at least some orange flags. Next, one assembles some evidence of risk or harm, quantitatively if possible, qualitatively where not possible. Then one looks at the costs of alternative proposals or remedial measures to do something about the harm that one perceives may be occurring. Given this model, if there is a plausible basis for believing that there is a significant threat due to the current activities by man and if the cost of remedial proposals don't portend serious economic hardship, one makes a political decision to act.

We in the environmental community think that the issue of acid deposition easily passes the test under this model of decisionmaking. We think it is the appropriate model of decisionmaking, given the difficulties that I have pointed out, both technical and ethical, associated with the use of cost-benefit analysis, and the pragmatic difficulties with the use of the more conservative ecosystem-conservative approach. We think it is just a matter of time before Congress acts on this matter.

In closing, I would have to agree with the comment by Curtis Moore, that the current debate is really about re-deciding what Congress already decided in 1970. In our view, in 1970 Congress decided to require that national ambient air quality standards be met and be met through the use of emission reduction techniques, not dispersion techniques, such as tall stacks and intermittent controls. In our view, if the 1970 act had been complied with rather than evaded, today's sulfur emissions would be anywhere from 6 to 10 million tons less than they are today. And now the period of evasion is coming to an end. The court has ordered EPA to adopt rules that finally do implement the requirements of the 1970 amendments for constant controls rather than dispersion. The Agency, early next year, is going to have to adopt final rules to implement that court order. Those rules will undoubtedly be litigated, but we believe a plausible scenario when that litigation comes to an end is that, in fact, compliance with the original Clean Air Act of 1970, as written, will produce reductions as great as 6 to 10 million tons from today's levels.

DISCUSSION

WARNER NORTH: Frankly, I was prepared to hear much more criticism from Mr. Hawkins with regard to the basic ideas of cost-benefit analysis and risk assessment. I have with me an editorial from the Washington Post from about a week ago which I was prepared to quote, and I think I still will. The editorial is on the subject of Mr. Ruckelshaus' departure from EPA, and it notes Ruckelshaus' contribution in promoting analysis as a means of determining where pollution abatement expenditures would do the most good. The editorial states, "Most of the environmental lobbies detest the idea of cost-benefit analysis, but it's much too useful a principle to neglect." It seems to me that there is a lot of merit in this judgment about cost-benefit analysis, and from Mr. Hawkins' comments I am not sure he disagrees. I heard him criticize some techniques of risk assessment and cost-benefit analysis, and I would agree with his criticisms. I myself was critical of EPA's original attempts to use probability assessment in the health risk assessment, in part because the judgmental probabilities were being averaged. I don't think that is good practice. The judgment of the scientists that underlies the probabilities should be stated and summarized. Discussion among the scientists may lead to a consensus on a probability assessment or clarification of the reasons why the scientists disagree in their probability assessments. This potentially useful interchange among the scientists will be lost if probability numbers are simply averaged together.

Similarly, I have problems with using wage indices to measure risk. I think that this usage may defeat the purpose we are trying to accomplish, for reasons that Mr. Hawkins articulated.

I am encouraged by Mr. Hawkins' comments, and I do not see that great a gulf between the decision model that he is talking about, his third category, and what I would propose to do under the labels of decision analysis and cost-benefit analysis. It's a question of using the analysis as a framework. This framework provides a means to bring together the various aspects of a complex environmental issue and set forth the judgments about these aspects explicitly. Each of us can then examine the judgments with respect to the information and with respect to the values, and decide for ourselves if we think the judgments are reasonable. In this fashion we may be able to either reach consensus or to have a better understanding of the basis for our disagreements.

RICHARD WILSON: On the whole I agree with that last comment. I would like to say, however, that when I mentioned that there is a difference in what people would call the value of the last lake and the value of the first lake, I was of course trying to emphasize the point that when there has been major ecological damage, people would want to take major steps to prevent further damage, while minor damage might be acceptable. There are two questions. First, can you get agreement on when that point comes? Mr. Hawkins and a lot of environmental groups say that it's now. But it is not obvious that everybody or even a majority says that. The other question, I think, is that before you get to that point, should you still do something to reduce damage, and how much do you decide to do? I believe you should do something, even a lot, to make sure that you don't even approach that point of major damage. The techniques of cost-benefit analysis and risk benefit analysis are very important. I regard what Mr. Hawkins is talking about as a cap on the whole procedure. Analytically I would say that the parameter alpha goes to infinity at a certain amount of damage.

DECISIONMAKING IN THE ABSENCE OF SCIENTIFIC CERTAINTY

SCIENTIFIC RESEARCH, RISK ASSESSMENT, AND POLICY DEVELOPMENT

Jay S. Jacobson

Plant Physiologist
Boyce Thompson Institute
Ithaca, New York 14853

INTRODUCTION

Assessing the risk of environmental hazards poses numerous difficulties. This paper attempts to explore some of these difficulties, presents reasons why they have existed, and offers some suggestions for improving the assessment process.

PROBLEMS ASSOCIATED WITH PREVIOUS ASSESSMENTS OF THE RISKS OF ENVIRONMENTAL HAZARDS

Decisionmakers are Dissatisfied With the Type of Information They Receive

Communication between scientists and decisionmakers always has been contentious. Galileo presented his version of the truth to existing powers in the 17th century and he met with an extreme reaction. When modern scientists speak "truth to power" and their "facts" are not in accord with expectations or needs, the reaction may be annoyance, disbelief, or a shift in research funds. Here is an example of a Congressman's response to recent expert testimony on the acid deposition issue.

"I must say I was disappointed in the testimony ... the double negatives were symbolic of the fuzziness, the queasiness and faint-heartedness of the testimony. Maybe on some other occasion, you'll speak a little bit more directly and without so much circumlocution and hedging ... and you'll give us a little more guidance than we've been able to get out of you this morning (U.S. House of Representatives, 1983)."

Politicians ask: "Who are the culprits? Where are they located? What effect on deposition of acidity would there be from a given reduction in emissions of precursors?"

Scientists answer that we know that sulfuric and nitric acids are deposited from the atmosphere. We know that sulfur and nitrogen oxides from combustion of fossil fuels are precursors of these acids. But the complexity of atmospheric physical and chemical processes is so great that simple answers to these questions cannot yet be given.

Politicians ask: "How much of the acidification that occurs in our soils, streams, and lakes is caused by deposition of acids from the atmosphere?"

Scientists answer that there is an important involvement of natural processes and so much variation among ecosystems that there is no single answer for all soils, streams, and lakes.

Politicians ask: "Are crops and forests damaged by acidic deposition and, if so, how severely?"

Scientists answer: "We're not sure if they are, nor what the mechanism is. There may be different answers for different crops and forests."

Politicians ask: "How much of a reduction in acidic deposition is needed in order to allow affected ecosystems to recover?"

Scientists answer: "We need to understand the whole chain of events from emissions through transport and transformations to deposition and effects in order to answer that question. Furthermore, we need to understand the basic processes by which ecosystems respond and recover from acidification. That will take a lot of additional research."

Politicians respond: "We've heard that answer before."

It may be that the initial questions are too simple for the complex nature of the problem. And perhaps expectations concerning the speed with which scientific research can answer these questions are unrealistic.

Assessment Staff Find That Scientific Evidence is Insufficient for Their Needs

Assessment staff have the difficult responsibility of compiling, molding, analyzing and presenting scientific information in a form that provides a base for making policy decisions. They choose or are given a particular assessment technique, such as economic valuation, and they try to fit the available information into the assessment mold. Inevitably there are gaps and discontinuities. Sometimes the scientific information must be stretched to fit places where it was not so intended by those who designed the experiments. Assessment staff find that they often do not have the kind, quantity, and quality of information they need to perform an adequate assessment.

Scientists are Frustrated by Inappropriate Uses of Their Data and Inconsistent Support for Research

Biological research can be subdivided into four categories (Figure 1), although any individual study may contain elements from more than one category. Exploratory research attempts to determine whether a phenomenon can occur. It may consist of observations or measurements, and its aim is to identify testable hypotheses. Confirmatory research attempts to prove alternative hypotheses. For example, it may establish quantitative relationships between some measure of pollutant dose and some type of plant response. Predictive research attempts to evaluate the factors that determine or modify the response in order to extrapolate from limited populations, time periods, and experimental conditions to actual circumstances. Finally, explanatory research attempts to determine the mechanism by which effects occur. For example, it may investigate the basis for susceptibility and tolerance to pollutants and the physiological and biochemical events leading to the occurrence of effects.

The listing of these categories does not imply that they always are performed in the sequence given in Figure 1. Ideally, scientists would like to understand the mechanism of action prior to prediction; however, this situation rarely occurs. Quantitative relationships between dose of toxic agent and effect on receptors usually are developed prior to explanations of how these relationships come about.

Certain kinds of assessments are appropriate and certain ones are inappropriate at each level of scientific knowledge. A complete and rigorous economic assessment, for example, would be based on knowledge obtained in all four research categories (Figure 1). A less complete and less rigorous but still useful economic assessment requires information from the first three categories. Qualitative risk assessments can be performed with information provided by the first two categories of research, but the economic costs of pollution cannot be estimated. If research investigations have not proceeded substantially beyond the first category, then not even a qualitative risk assessment can be performed.

We can use experiments with simulated acidic rain and agricultural crops as an example. To extrapolate from experiments conducted with field-grown crops grown in one field in one summer to a regional assessment it is necessary to know the effects of soil and climatic variations and differences in cultivation conditions on crop response to acidic rain. It is also necessary to know how differences between simulated and actual acidic rain exposures may affect plant response. Where we only have information from studies performed in the greenhouse, we also would like to know how differences between greenhouse-grown and field-grown plants affect the dose-response relationships. It is important to identify the point in the development of information concerning crop response and acidic rain at which economic assessment is justified.

Category of Research	Category of Policy Analysis
Exploratory (develop hypotheses)	Not Feasible
Confirmatory (test hypotheses)	Qualitative
Predictive (quantitative relationships, extrapolation)	Quantitative
Explanatory (mechanism of action)	Quantitative

Figure 1: Approaches to Biological Research and Relationship to Policy Analysis

A final difficulty concerns support for research on environmental issues. Acid rain is a problem that has been occurring for many decades and will, no doubt, continue to occur into the next century. We are concerned about effects on ecosystem processes that have been in existence for millions of years. We see that, at every stage of the problem, from emissions to transport and transformation, to deposition, and to effects on ecosystems, there is intimate involvement with natural processes. We can give reliable guidance for assessments of acid rain only to the extent that we understand these processes. Yet we fail to see the connection between improvements in our knowledge of fundamental processes and advances in understanding of acid rain.

SOME REASONS FOR PROBLEMS WITH ASSESSMENT OF ENVIRONMENTAL HAZARDS

What are the essential requirements of scientific evidence if the evidence is to be useful as a basis for developing national policies (Figure 2)? From the point of view of decisionmakers, there is one requirement that seems to be most important, namely, the need for agreement among experts on the cause and extent of the problem and its solution. When there is disagreement and no single, clear interpretation of scientific evidence, decisionmakers find it difficult to identify and obtain support for a particular strategy.

Economists have several requirements for scientific evidence that forms the basis of an assessment. Although in principle economists can deal with evidence that is not quantitative, confidence in conclusions concerning, for example, the cost of acidic deposition to agriculture, require quantitative estimates of the effects of known doses of acidity on the yield and quality of marketed commodities. This information is needed for the existing range of doses of acidic deposition and for increases and decreases in acid deposition that might result from the imposition of controls on emissions of precursors of acidic deposition.

The economist engaged in assessment activities usually is more interested in breadth of information and fidelity to actual conditions. The biological scientist, however, usually has quite different concerns. Biologists are interested in depth of knowledge and certainty of conclusions. They attempt to provide unambiguous demonstrations of cause and effect. Rarely is the range of information needed by economists provided unless the research program initially is designed to include a

Criteria of Decisionmakers

Agreement among experts, clear relationship to policy

Criteria of Economists

Relevance to actual conditions, breadth of information

Criteria of Scientists

Certainty of cause and effect, precision of information

Figure 2: Requirements of Scientific Information

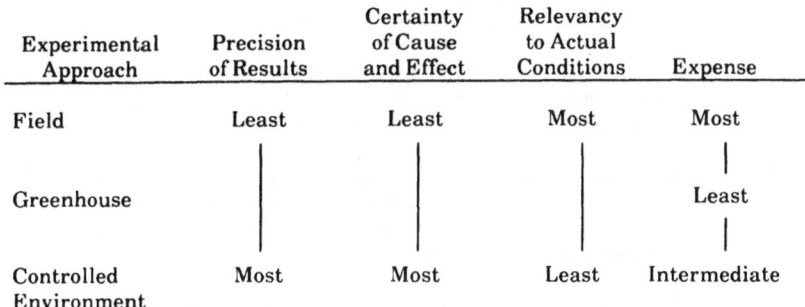

Experimental Approach	Precision of Results	Certainty of Cause and Effect	Relevancy to Actual Conditions	Expense
Field	Least	Least	Most	Most
Greenhouse				Least
Controlled Environment	Most	Most	Least	Intermediate

Figure 3: Characteristics of Experimental Approaches to the Study of Effects of Pollutants on Agricultural Crops

modelling component. Experimental scientists attempt to minimize variability by selecting conditions and by controlling variables that inevitably divorce the experimental situation from actual conditions (Figure 3). There usually is an inverse relationship between ability to relate cause with effect and the relevance of results to ambient conditions.

The different approaches and requirements of scientists, assessment personnel, and decisionmakers are among the most important reasons for past problems with the assessment process. Another explanation concerns the lack of integration among these three building stones of an assessment.

Sound scientific research may be quite useless for risk assessment and policy development. Assessment techniques may not fit the needs of decisionmakers nor make appropriate use of the available information. Perceived policy needs may be unrealistic given the available knowledge and techniques of assessment. The needs of decisionmakers, the requirements of assessment techniques, and the methods of experimental research must be integrated. Each leg of a three-legged chair may be sturdy, but if they are different lengths, the chair will be unstable. In our haste to produce immediate answers, we ignore this integration and foster conflict, extreme reactions, and paralysis of the decisionmaking process, all of which characterize the acid deposition issue in the U.S.

WHAT CAN BE DONE TO IMPROVE THE ASSESSMENT PROCESS?

The first step in improving the assessment process concerns identifying the criteria to be used for evaluating environmental impact and for determining whether action should be taken. If we have not established the criteria for different levels of action, then, even with new information and resolution of the scientific issues, we still will be unable to make decisions. We will have emphasized science and assessment and will have ignored the basis on which policy decisions are to be made. Some simple examples of alternative criteria for decisions are presented in Figure 4.

I. Take action only if the monetary cost of acid rain is greater than the cost of emission controls (cost-benefit criterion).

II. Take action only if the harmful effects of acid rain are greater than the harmful effects of other pollutants such as ozone (most-damaging pollutant criterion).

III. Take action only if the harmful effects of acid rain are greater than the effects of natural factors such as insects, diseases, climate, biological and geological processes (natural phenomena as criteria).

IV. Use none of the above as criteria for action. Take action on the basis of respect for the desires of neighboring states and nations and considerations for future generations (the good-neighbor policy and the intergenerational issue as criteria).

Figure 4: Alternative Criteria for Taking Action on Acid Rain

If we decide that the cost-benefit approach is the best way to make decisions concerning acidic deposition, then we should select the appropriate assessment techniques and plan research programs that will supply information for this particular kind of assessment. If we decide that comparisons of the effects of acidic deposition with the effects of other pollutants provides the appropriate criterion for action, then we would perform different kinds of experimental research and modify the assessment technique. If our decision is based on comparisons of the effects of acidic deposition with natural processes, then the assessment techniques and scientific investigations must be chosen specifically for this purpose. Using the criteria of the good-neighbor policy and concern for future generations, we would approach assessments and research quite differently.

This first step leads to a second suggestion for improving the assessment of environmental hazards. We should recognize that assessment is an evolving process that must allow for improvements in methodology and even qualitative shifts in the nature of assessment techniques as new evidence is provided. Perhaps we should use one decision criterion at an early stage of knowledge and use different or additional decision criteria at later, more advanced stages of knowledge. The growth and evolution of each leg of the triad (science, assessment, and decisionmaking) should be nurtured so that a reasonably coherent product is delivered as the depth and breadth of knowledge improves.

REFERENCES

U.S. House of Representatives, 1983, Hearings before the Subcommittee on Energy Development and Applications and the Subcommittee on Natural Resources, Agriculture Research and Environment of the Committee on Science and Technology, 98th Congress, First Session, September, 1983. USGPO., Washington, 1287 pages.

RESPONSE TO JAY JACOBSON

Rosina M. Bierbaum

Assistant Project Director
Office of Technology Assessment*
U.S. Congress
Washington, D.C. 20515

Several participants in this conference suggested that it would be useful if I were to discuss the current flurry of acid rain bills in Congress in light of the papers in this session. Dr. Jacobson described the difficulty of integrating available scientific information into viable political options and the frustration both scientists and politicians feel. Scientists try to ascertain and deal with certainty. Though no one is truly comfortable, politicians are more comfortable than most addressing uncertainty and dealing with risks. The scientist pales at the scope of the acid rain problem in its entirety and at attempts to crudely assess the magnitude of all potential resource damage to lakes, forests, streams, crops, health, visibility and man-made materials.

Ideally, we all would like a curve such as shown in Figure 1, a well-understood relationship between pollution control expenditures and potential resource damage. Each point on the curve represents a given amount of resource damage prevented. Benefit-cost analysis considers expenditures to be worthwhile until a point is reached where additional control costs exceed benefits from the resource damage prevented.

Though the benefit-cost test is useful for considering many policy questions, we have seen during the course of this conference that the assumptions underlying the analysis and the attributes of the transported air pollution problem are currently poorly matched. First, the relationship between emissions and resource damage is not well enough defined. Second, many of the resources of concern are not easily assigned monetary values. Finally, the test does not consider the distribution of costs and benefits. Instead of our desired curve, we have the shaded box in Figure 1. The corners of the box illustrate the complete range of possible outcomes: the effects of the range of available control choices (emission control costs, mitigation costs, etc.), and the range of resource effects (potential damage to lakes and streams, forests and soils, materials, visibility and human health).

Consider the left-hand side of the diagram: little or no pollution control expenditures beyond those called for under existing regulations. The range of outcomes, characterized by corners "a" and "b", represents the uncertainties involved in the magnitude of resource damage associated with a given level of transported air

*The views expressed in this paper are entirely those of the author and not necessarily those of OTA.

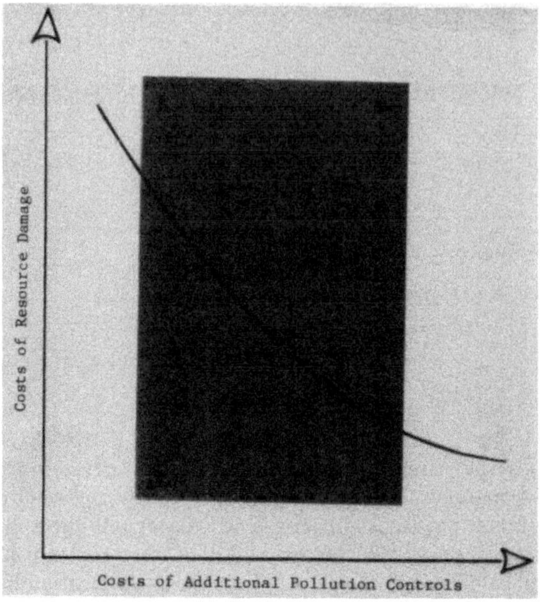

Figure 1.

pollutants. The right-hand side of the diagram, corners "c" and "d", shows a similar range of outcomes associated with high pollution control expenditures. This range illustrates the uncertainty about the effectiveness of a chosen control strategy.

Not all points within the shaded box represent equally likely outcomes. Resource damage is more likely to occur with current or increased levels of emissions than with reduced levels; but estimates of current damage, or of how effective pollution control programs might be in preventing potential damage, are difficult to obtain. All we can do now is bound the range of potential outcomes. The risks of the control costs can be described more accurately than the risks of resource damage; thus, the gray band is shown as narrower for the cost axis.

Why then, have 20 bills to address acid deposition now been introduced in the 98th Congress? Sixteen of these seek 8-12 million ton reductions in sulfur dioxide (about 35-50 percent). (This is much more than the 6-million ton reduction level that we were told yesterday might be considered "cost-effective".) One in the House, H.R. 3400, the Sikorski/Waxman bill, has 128 co-sponsors and S. 768 was passed by the full Senate Committee on Environment and Public Works.

What goes on in a policymaker's mind? How does the policymaker use the science that's available and circumvent the unknowns? How were the bills under consideration designed?

The first thing I was asked to do when I began as a Congressional Fellow at the Office of Technology Assessment after 10 years of college and graduate training in science, was to draw a flowchart of how transported air pollutants should be studied to provide Congress with an estimate of the risks of acting now to control transported air

LRTAP Processes

Sources

Utility * Industrial * Mobile * Residential * Natural

Atmospherics

Acid Deposition * Sulfur and Nitrogen Oxides * Ozone * Particulates

Resources

Lakes * Forests * Crops * Groundwater Materials * Health

Figure 2.

pollutants versus the risks of waiting. As a scientist, without hesitation, I drew the following (Figure 2). (This figure is actually quite similar to the process of the National Acid Precipitation Plan.) One should look at the current and historical patterns of emissions, atmospheric transport and transformation, deposition patterns and natural resource effects, and the policy would follow logically from the science. It was a linear process. To a scientist, the level of knowledge drives the higher concerns. Science is truth, right?

Most members of Congress do not primarily care about the science of acid rain; the concerns of their constituents are paramount. Their flowchart looks more like a series of Chinese boxes (Figure 3). Decisions are made about national energy policy, pollution control policies, natural resource policy and agricultural policy. Societal

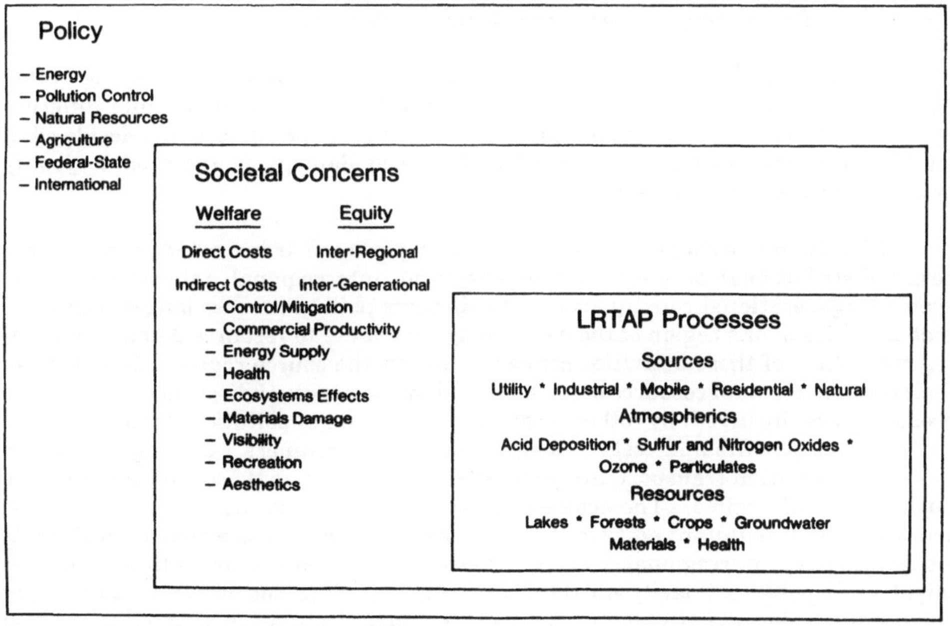

Figure 3. Long-Range Transport Air Pollutant Study Approach

concerns about equity and welfare temper these policies. Where appropriate, science, the underpinning, is brought into play.

THE CONGRESSIONAL DILEMMA

What then, is the Congressional dilemma regarding transported air pollutants? It is threefold. It is a conflict of science, but it is also a conflict of values, and of distributional issues.

Four years have passed since the introduction of the first acid rain control bill and the initiation of the 10 year National Plan. Debate over scientific understanding of transported air pollutants remains perhaps the most visible aspect of the policy controversy. Many assert that the causes and consequences of acid deposition are both sufficiently well understood and significant to warrant immediate action to control it. Others emphasize the complexity of the phenomena involved, and argue that until more is known about the benefits of control, no regulatory strategy can be justified on the basis of existing scientific knowledge. Scientific uncertainty is not new to air pollution policy. The continuing controversy over the nature and magnitude of health risks, the critical measure for setting national ambient air quality standards, is perhaps the most obvious example of the difficulty of unambiguously documenting the scientific basis for regulation.

Disagreements over facts about transported air pollutants are accompanied by disagreements over values. Even if a scientific consensus existed on the magnitude of the problem, policy choices would still be complicated by lack of agreement over how to promote economic development while protecting the environment. While the concept of a tradeoff between these two values is widely accepted, various individuals and groups differ sharply on where the balance should be struck.

Disagreements over facts and values are intertwined. Differing value structures among various individuals and groups lead each to draw quite different conclusions from the same body of scientific information, and subjectively decide what level of scientific certainty and/or what degree of damage is required to justify undertaking a control program.

Finally, and perhaps of greater political importance, transported air pollutants pose a distributional problem with intersectoral, interregional, international, and even intergenerational consequences. Winds carry pollutants over long distances so that activities in one region of the nation may contribute to resource damage in other regions. Many of these activities primarily benefit the source region, while some of their costs, in terms of resource damage caused by the eventual disposal of their waste products, are incurred by other regions. Long-range pollution transport thus redistributes benefits and costs among regions. The "winners," if Congress acts to control air pollution transport, are generally not the same as those who stand to lose under a control regime. The ecologically sensitive areas do not correspond to the areas of highest pollutant emissions. Programs to control transported air pollutants would also have interregional distribution aspects. The costs of controlling emissions might be imposed primarily on the source region, while the major benefit might accrue to the downwind receptor region. We have a situation that pits the East against the Midwest, the U.S. against Canada, industry against recreation, the next generation's resources against our economic well-being.

KEY SCIENTIFIC UNCERTAINTIES

Keeping in mind that values and distributional issues affect the use of available technical information, let us look at the five key scientific uncertainties that have been the focus of the policy debate. These include controversies about (1) the extent and location of current damages, (2) future damages, and whether they are cumulative and/or irreversible, (3) the geographic origins of observed levels of pollution, (4) the effectiveness of emissions reductions for reducing observed levels of transported pollutants, and (5) whether a research program will provide significant new results. These scientific uncertainties have significance for issues of values and distribution discussed earlier, including (1) making air pollution control policy as fair as possible, (2) considering intergenerational effects of action or nonaction, (3) weighing the risk of damage against gains that might be achieved by waiting for better information or technology, (4) assuring that the benefits of a control program, in the broadest sense, justify the cost.

Uncertainty About the Extent and Location of Current Damages

Effects researchers are certain that transported air pollutants have caused some damages. At issue is the extent of the damage, whether it is fairly localized or widespread, and which resources are affected. For example, there is little question that ozone damages crops, and that acid deposition damages lakes. We can regionally characterize the areas where these resources are at risk.

Figure 4 shows ambient ozone concentrations across the country. We can see that ozone levels are high in much of the crop-producing regions of the country. Figure 5 shows (in gray) the areas of the country in which we would expect sensitive

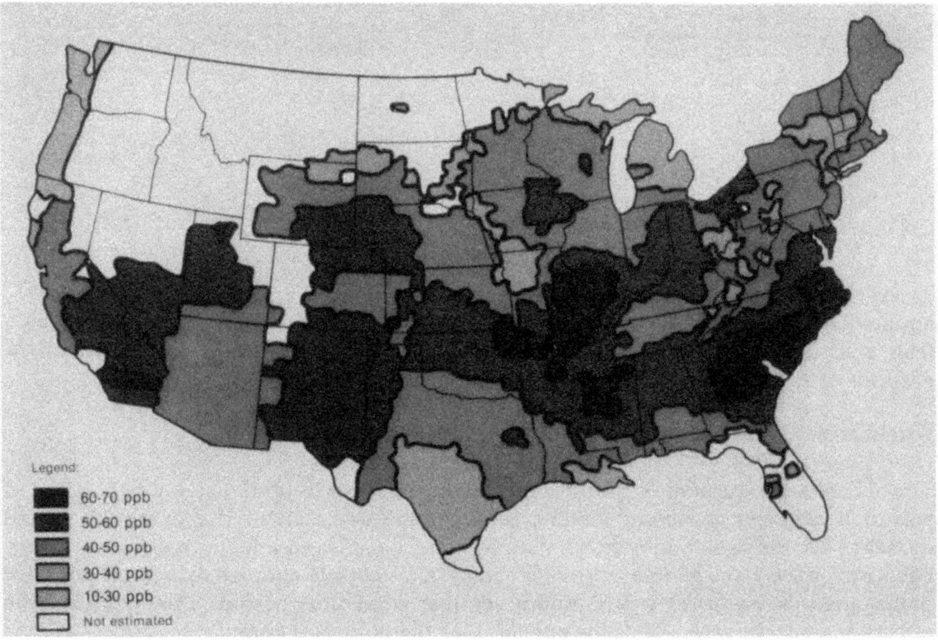

Figure 4. Ozone Concentration: Daytime Average for Summer 1978

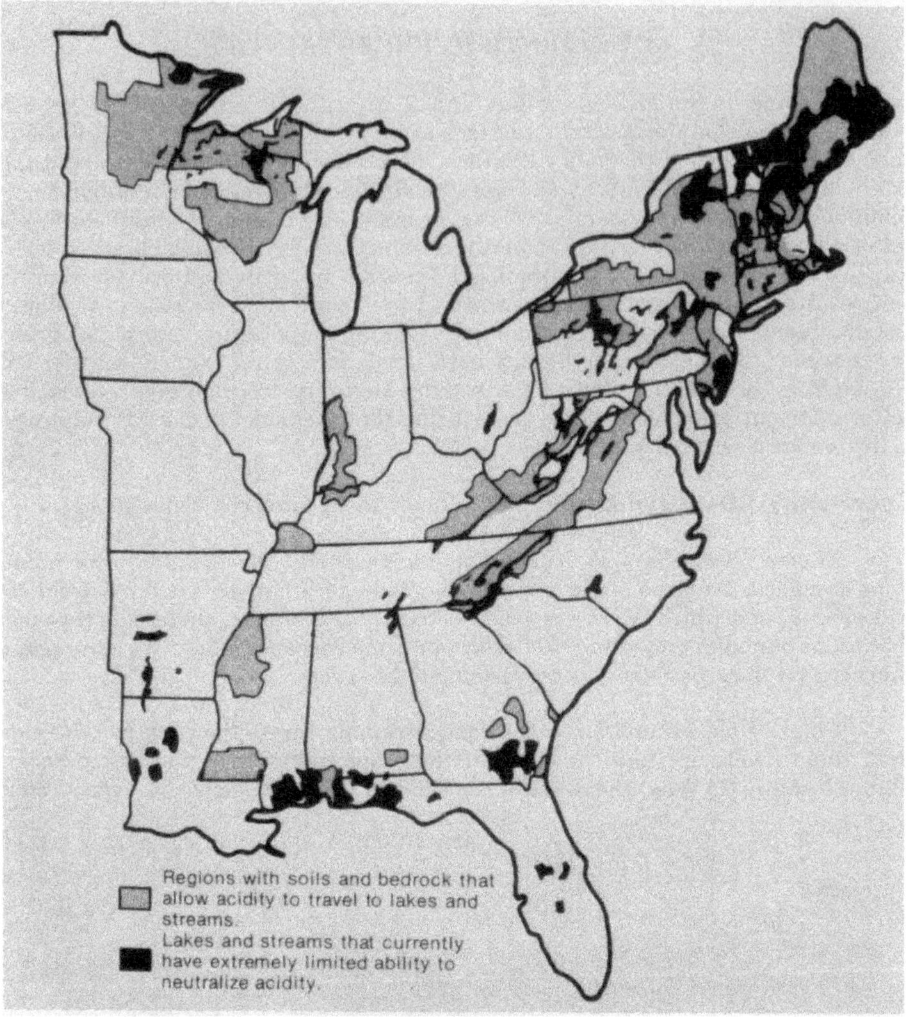

Figure 5. Regions of the Eastern United States With Surface Waters Sensitive to Acid Deposition

aquatic resources to be located. For other concerns, for example, damage to forests from acid deposition and ozone, the uncertainties are so large that it is difficult to describe even the patterns or magnitude of the risk.

Uncertainty About Future Damage

For some resources, pollution-related damages might worsen over time if pollution remains at about current levels. The question of the extent to which damages accumulate, and over what time scale, has played a major role in the policy debate over the risks of delaying control action. A closely coupled concern is whether damage can be reversed easily, and if so, over what time period. For example, the effects of a severe series of ozone episodes, or the potential cumulative effects of acid deposition, will persist in the forest community until a new forest grows. Uncertainties about the irreversibility of resource damage make delaying action until better information is obtained highly controversial. To the extent that damages are both significantly cumulative and irreversible, waiting for better information would cause resources to be irretrievably lost for this generation and the next several to follow.

Uncertainty About the Origin of Acid Deposition

Parts of this question have been substantially answered since 1980. Though pollutants of natural origin cause some acid deposition, deposition over large areas of the eastern United States far exceeds the level attributable to natural sources of pollutants. In addition, while local sources (within 30 miles) do contribute to acid deposition, most analyses indicate that a large share of the deposition, in some cases, over half, originates from both medium-range and distant sources (greater than 300 miles away) as well.

Figure 6 shows the average distance a sulfur molecule travels. We can see that in areas of high emissions, such as the Midwest, the average is less than 180 miles. In the Northeast, the average sulfur molecule deposited has traveled much farther. The remaining key uncertainty in this area is whether scientists can reliably determine how much of the deposition in any one region originates from emissions in any other.

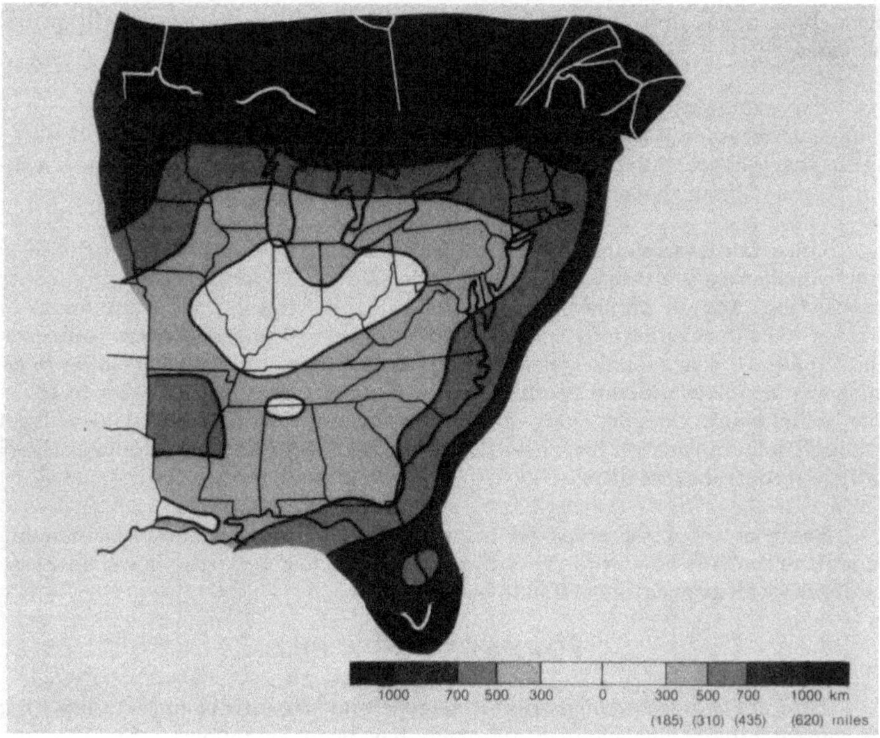

Figure 6. Average Sulfur Transport Distances Across Eastern North America

Computer models of varying sophistication are available to perform such analyses, but are most useful in delineating broad regional patterns on seasonal timescales. The inherent variability of weather and the complexity of atmospheric chemistry make it unlikely that models will ever be able to predict how much one individual source contributes to deposition in a small area.

Uncertainty About the Effectiveness of Emissions Reductions for Reducing Acid Deposition

Several analyses conducted in the last four years have also helped to answer this question. They indicate that cutting back regional sulfur dioxide emissions alone will significantly, but not quite equally, reduce the amount of sulfur deposited in various forms. Because other pollutants such as hydrocarbons and nitrogen oxides can enhance or impede chemical transformation of sulfur dioxide, simultaneous control of other pollutants might better achieve the policy objective of reaping the greatest possible benefit from the costs of controlling emissions. But how such a multiple-pollutant control strategy should be structured will not be known for many years.

Uncertainty About the Results of the Research Program

One of the most difficult decisions facing the Congress is whether to act during this reauthorization of the Clean Air Act, or wait for results of the ongoing National Plan. While the research efforts are intended to reduce the uncertainty discussed above, how much new insight 5 to 10 years of further research will provide is unknown.

For example, years to decades are required to observe changes in many ecological processes. Patterns of crop yield and forest productivity typically vary from year to year; separating the effect of acid deposition and ozone from normally expected year-to-year fluctuations requires many years of data.

Uncertainties about the progress of research programs are an important factor when considering the timetable for policy decisions. Given the planning, contracts, construction, and so on necessary to significantly reduce pollutant emissions, a decision to control emissions now may still require 10 or more years to implement. Waiting 5 to 10 years for the results of a research program before deciding to control emissions increases the time required to reduce deposition from 10 years to 15 or 20, a time scale many consider long enough to be significant to ecological systems. Although delaying control increases the risk of resource damage, it reduces the risk of inefficient control expenditures.

Some of the bills proposed in the 98th Congress seek to avoid delays by mandating controls now, while retaining the option to change the law if new research results show an alternative action to be preferable.

FORMULATING A BILL

How did Congress use available science and circumvent uncertainty to design the control bills? The generic approach has been as follows. Six fundamental decisions were to be made.

1. Which Pollutants to Control

In the eastern U.S., sulfur compounds are responsible for about two times more acidity in precipitation than are nitrogen compounds. Sulfur compounds are known to be related to freshwater acidification, visibility degradation and materials damage, and are implicated in health effects. Since sulfur in fuel delivered is measured, sulfur dioxide emissions inventories are much better developed than nitrogen oxide inventories, making reductions more determinable. So sulfur dioxide was the logical choice for initial control. Some bills, recognizing that nitrogen oxides may also be a problem, seek to freeze emissions or achieve a small reduction.

2. Where to Control

Several regions are potential candidates. Figure 7 shows the pH of rainfall; a small number of states within the "bull's-eye" of highest deposition up to the 48 contiguous states could be identified for control.

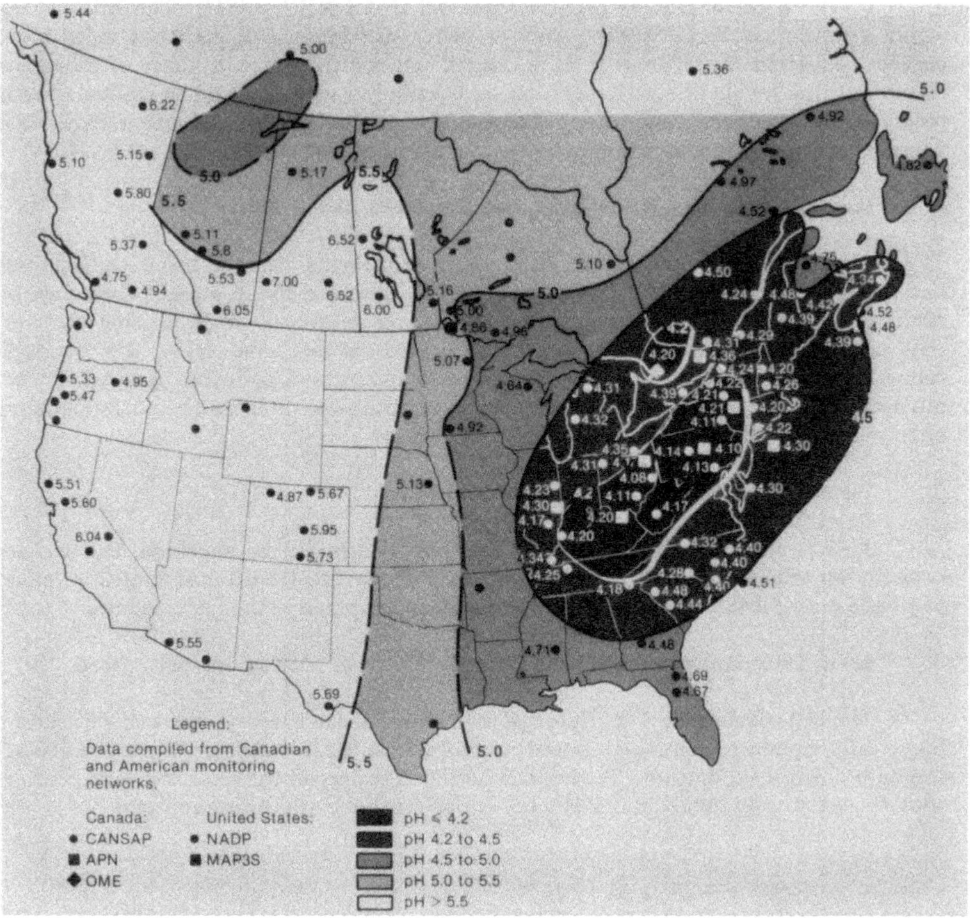

Figure 7. Precipitation Acidity -- Annual Average pH for 1980

All of the bills to date identify either a 31-state region, which includes all states east of and bordering on the Mississippi River, or the continental United States, thus implicitly defining acid deposition as a regional problem requiring a regional solution.

3. How Much to Reduce

Two types of strategies are available for choosing a socially-desirable level of emissions reductions: (1) decide how much to spend, and control up to a level for which resources are available, or (2) define a level of resource protection, and spend the requisite dollars necessary to achieve that protection. For example, if Congress had decided that one billion dollars was an appropriate expenditure given current uncertainties, perhaps 2 to 5 million tons of sulfur dioxide could have been removed for that amount. However, no bill used this approach. Concerns for aquatic resources drove the bills to an environmental target approach. While levels of control affording resource protection are difficult to determine, there is some discussion of the matter. A report of the National Academy of Science (NAS) has reported that no lakes are acid in areas where pH of rain is greater than 4.6 (NAS-NRC, 1981). The bills which would eliminate 8 to 12 million tons of sulfur dioxide emissions, that is, a 35-50 percent reduction in the eastern U.S., might achieve this goal in many of the areas receiving high levels of acid deposition, but probably would not do so in those areas receiving the greatest amounts. Figure 8 shows the percent reduction in sulfate necessary to get the pH of rain up to 4.6.

4. How to Allocate Emissions Reduction

Allocation formulae can take many different forms. In fact, the 97th Congress had quite a "formula war". Emissions could be reduced by an equal percentage in each state regardless of whether the state is a relatively high or low emitter. Alternatively, allowable emissions rates could be set, requiring the greatest reductions from sources that emit the greatest amounts of pollution per unit of fuel burned. This generally less costly approach of emission rates has been the primary basis of all bills introduced this session of Congress.

5. Who Should Pay for the Reductions

This is clearly a distributional question. As might be expected, the answer depends on who introduced a particular bill. Both the traditional "polluter pays" principle and efforts to distribute costs have been proposed in the various bills.

6. How Can Undesirable Effects of a Control Policy be Mitigated

The bills espousing a "polluter pays" principle don't worry about this question. Some bills mandate technology-based control measures in whole or in part to protect the high-sulfur coal market. Taxing mechanisms to spread the costs to a larger group than those actually involved in reducing emissions have also been proposed.

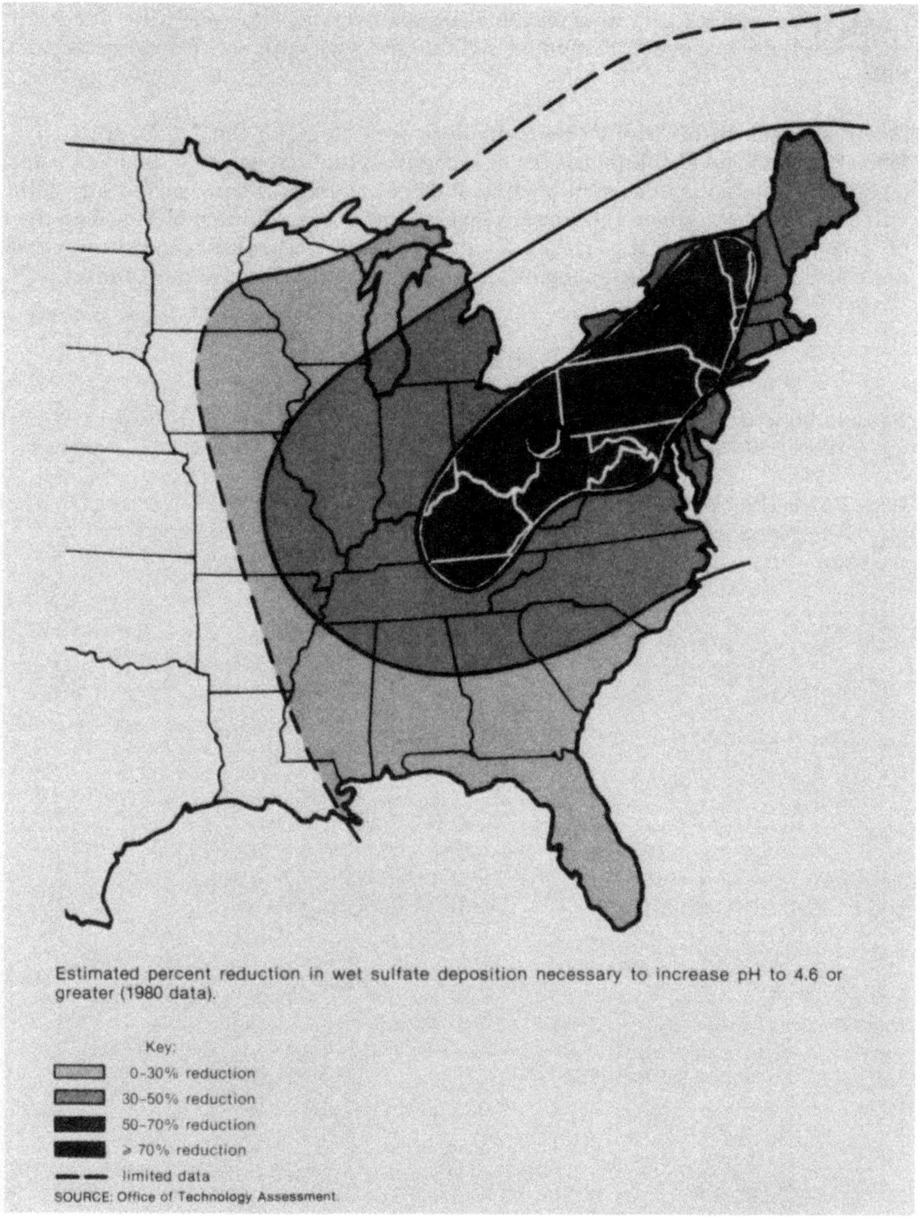

Figure 8. Target Value: Average Precipitation pH of 4.6

CONCLUSION

Those who answered these six questions have designed control bills. In these cases, value choices and distributional issues overrode the scientific uncertainties. However, there are over 200 members of Congress who did not co-sponsor any of these bills.

Scientists seek to be "right"; policymakers seek to make the "right decision". The former refers to an absolute; the latter recognizes that not acting is as much a decision as acting. The policy decision to control or not to control transported air pollutants will be made now. Given the current level of knowledge, it must be based on the risks of resource damage, the risks of unwarranted control expenditures, and the distribution of these risks among different groups and regions of the country.

REFERENCES

National Research Council, "Atmosphere-Biosphere Interactions: Towards a Better Understanding Ecological Consequences of Fossil Fuel Combustion," National Academy Press, Washington, D.C., 1981.

Office of Technology Assessment, Congress of the United States, June, 1984, *Acid Rain and Transported Air Pollutants: Implications for Public Policy*. GPO Stock No. 052-003-00956-1.

ECONOMIC ANALYSIS: THE NATIONAL ACID PRECIPITATION ASSESSMENT PROGRAM*

J. William Currie
Ronald J. Moe
Ronald J. Nesse**

Battelle Pacific Northwest Laboratories
Box 999
Richland, WA 99352

INTRODUCTION

The conversion of acid precursors into acid deposition, as well as the relationship between acid deposition and its resulting effects on the environment, are complex and not well understood. The scientific uncertainties would be of little concern to policymakers were it not for two additional issues. On the one hand, the costs of reducing acid precursors and acid deposition appear to be substantial and of major concern to important segments in our society. However, the benefits of reducing emissions may also be large and are also of interest to our society. To many, both economists and policy analysts, controlling acid deposition becomes an evaluation of the gains and losses of any proposed action. While economists often think about these trade-offs in monetary terms, policy analysts often do not.

Recognizing these scientific and political uncertainties, the National Acid Precipitation Assessment Program (NAPAP) was implemented in 1980 with two primary objectives. First, it seeks to fill many of the gaps in scientific understanding of the acid deposition phenomenon through a comprehensive research program. Second, NAPAP seeks to assess the consequences of various methods of reducing acid deposition. These consequences are to be measured both in physical changes in the environment and in economic terms.

The unanswered questions about the causes, extent, severity, and solutions to the ecological hazards posed by acid precipitation far outweigh our present understanding of the phenomenon. The comprehensive research program undertaken

*This research has been funded as part of the National Acid Precipitation Assessment Program by the Environmental Protection Agency under Related Services Agreement with the U.S. Department of Energy under Contract DE-ACO6-76RLO 1830.

**J.W. Currie, Associate Manager, Energy Systems Department; R.J. Moe, Senior Research Economist; and R.J. Nesse, Senior Research Economist.

by the federal government seeks to provide an improved scientific basis to permit the development of a balanced, cost-effective, and ecologically sound national response to this problem.

Concerns With Respect to Acid Deposition Control Policies

Society and policymakers are concerned with acid deposition and other long-range air pollutants because of the potential damages to the ecosystem and human health. However, they are also concerned that controlling such pollutants could result in costs significantly exceeding any damages. The most commonly expressed concerns with respect to acid deposition and control policies are listed below.

- Acid deposition alone or in combination with other pollutants is known or suspected of killing fish, harming forests, and damaging materials. Individuals are concerned with the effects on these goods and resources since they directly affect our well-being. In addition, individuals who do not use the resources are concerned with species preservation or maintenance of ecosystems for future generations, referred to as non-user or "intrinsic" values.

- The effects of acid deposition policies are difficult to predict. In addition to mitigating the effects of acid deposition, most control policies under consideration would also improve visibility and possibly human health by reducing the pollutants that cause acid deposition. Studies estimating the value of increased visibility have found it to be substantial. These ancillary benefits which result from policies designed to reduce acid deposition, must be factored into the discussion of whether and how to control acid deposition.

- Controlling acid deposition will be expensive. The Office of Technology Assessment estimates reducing sulfur dioxide emissions by 10 million tons will cost $3 to $4 billion annually (OTA 1984, p. 14).

- The cost of reducing pollution could affect local jobs and industries. If utilities reduce sulfur dioxide emissions by switching to a lower sulfur coal, jobs will be lost in high-sulfur coal areas.

- Electricity price increases may not be uniform and, although small, could affect certain high-users of electricity. For politicians who must rely on local support for election, such problems become important concerns in the acid deposition debate.

While NAPAP emphasizes basic scientific research, it is also responsible for addressing the policy issues. The economics component of NAPAP is designed to deal with these policy concerns.

ECONOMICS AND THE ACID DEPOSITION PHENOMENON

The National Acid Precipitation Program is currently funding economic research to generate information required to prepare assessments in 1985, 1987, and

1989 (see Appendix). Below we discuss the information required by decisionmakers, the economic modeling capabilities required to satisfy these informational needs, and the analytical techniques available to develop such modeling capabilities. Finally, we evaluate the suitability of these techniques for developing the required modeling capabilities and the usefulness of the information that can be generated using models based on available techniques.

The presence of economic research in NAPAP is due in large part to the kinds of information that Congress has specified will be included in the assessments. In general, these assessments must provide information that decisionmakers in the executive and legislative branches of the federal government, and others, can use to make informed policy decisions regarding acid deposition. To determine how important the acid deposition problem is, these decisionmakers require information about current and expected future impacts of acid deposition on their constituents and the nation. To determine the appropriateness of a particular policy designed to alleviate the problem, many decisionmakers desire information about the total beneficial and negative impacts that such a policy will have on their constituents and the nation as a whole.

These general information requirements include economic impacts. Thus decisionmakers require accurate, defensible estimates of the current and expected future monetary damages caused by acid deposition in the absence of control policies. They also require accurate, defensible estimates of the total national current and expected future monetary benefits and monetary costs of specific reduction or mitigation policies. In addition to total national estimates, decisionmakers require estimates of impacts on specific components of the population, e.g., a particular region or income class. They also require information about the level of certainty that can be attached to such estimates.

NAPAP has determined that the assessments are not to be only an accounting of benefits and costs, but will also include a great deal of scientific and policy information. For example, they will include estimates of the physical damages of acid deposition, such as changes in fishable acres, and information on the impacts of particular policies in nonmonetary terms, such as estimates of emissions and deposition levels. This information will be presented in the assessment in the same way monetary values are presented.

The economic research program focuses on developing models that will be capable of producing the economic information required for these assessments. Basically, the information requirements lead to the following desired characteristics of the economic research program and individual models.

Exhaustive. Different decisionmakers desire different information about the impacts of their actions; to the extent possible, benefit models should be developed to value all of the potential physical and/or biological effects of deposition reductions, and cost models should be developed to value all of the inputs required to implement potential strategies.

Accurate and Defensible. Decisionmakers must have confidence that benefit and cost estimates are accurate and defensible; otherwise, they will not use the estimates. Models should thus be consistent with economic theory and based on analytical techniques and data that are accepted both by the profession and by intelligent laymen.

National. Since benefit and cost estimates are required for the entire nation, models should be developed that can generate estimates for all states or regions.

Current and Future. Since decisionmakers require estimates of both current and expected future benefits and costs, the models must be able to forecast future benefits and costs, using inputs generated by NAPAP or other available sources.

Linkages. Decisionmakers require information about the monetary values of specific physical and biological effects, such as fish deaths avoided, (associated with deposition reduction), as well as of the specific inputs, such as scrubbers, associated with emission reduction or mitigation strategies. Benefit models should, therefore, to the extent possible, use input estimates provided by NAPAP physical and biological models, and cost/emission models should provide outputs required by the transportation/transformation models. These linkages result in a consistent analysis and allow decisionmakers to extract information at the desired step of the assessment.

Uncertainty. Decisionmakers want to know the level of certainty or uncertainty that can be attached to specific benefit and cost estimates. Benefit and cost models should thus, to the extent possible, provide probabilistic estimates or ranges of values. Models should consider both the uncertainty in the economic process under study as well as uncertainty in the inputs from other NAPAP models.

Distribution. Decisionmakers require information about the distribution of policy impacts, including costs and benefits, across particular subgroups of the population. Models should, to the extent possible, estimate costs and benefits to particular regions and income classes.

Nonmonetary Economic Impacts. Decisionmakers require information about policy impacts that are economic in nature but cannot be easily valued in monetary terms. Changes in employment and income are particularly important, especially on a regional basis. Models to consider such impacts should be developed.

Policy Analysis Process. Policies to be considered in the assessments may not be known until shortly before the assessment is to be prepared, and the models developed as part of NAPAP may also be used for other policy analysis exercises. Therefore, models should be developed that can provide results in a relatively short time period at a relatively modest cost.

ANALYTICAL TECHNIQUES AVAILABLE TO DEVELOP COST AND BENEFIT MODELS

A number of analytical techniques exist that may form the basis for the modeling capabilities described above. These techniques may be broadly divided into two categories, those that consider the emissions or cost component of the acid deposition phenomenon and those that consider the effects of control strategies or benefits component.

Techniques for Cost Model Development

On the emissions or cost side, models are required that can forecast emissions of acid deposition precursors such as sulfur dioxide, and the capital and operating costs associated with a specific emissions control strategy. These models must be able to assess how particular policies affect each of the primary emitting sectors, the electric utility, industrial, transportation, and the commercial sectors, as well as how they affect the precursors of acid deposition, sulfur oxides and nitrogen oxides. The models must be capable of accounting for costs and emissions associated with specific control equipment such as flue gas desulfurization equipment ("scrubbers"), which future legislation or regulation may require emitters to install. The models must also be capable of predicting the costs and emissions associated with more flexible legislation or regulation, for example, the imposition of state emission ceilings.

The analytical techniques required to construct such capabilities are well developed, conceptually straight-forward, and reasonably well accepted. Economic-engineering models of the behavior of each primary emitting sector have been developed. These models forecast emissions of the pollutants of interest given forecasts of sector output or activity and information about the specific pieces of emission control equipment that have been installed; they also calculate the capital and operating costs associated with this equipment. Furthermore, these models are capable of identifying "least-cost" strategies for meeting specific emission levels; for example, the models would calculate the cost borne by a particular utility if it is told to reduce its annual SO_2 emissions at a particular plant by 20 percent, using any method it desires. These models can also identify "least-cost" strategies for meeting specific emission levels at either the state or national level, both for a single sector and across a number of sectors.

These models are basically engineering models that use economic information. The NAPAP model of the electric utility sector, for example, accepts forecasts of annual, state-level electricity consumption, fuel prices, interest rates, capital equipment prices for generating capacity and emissions control, and the results of the present and future emissions policies. Given these forecasts, the model identifies the least-cost method of generating the required electricity that also satisfies the emissions requirements, using general engineering principles.

Techniques for Benefit Model Development

Acid deposition affects a number of human activities, possibly including agriculture, commercial forestry, recreational fishing, the use of specific materials, and the recreational and aesthetic use of ecosystems in general. It affects these activities by changing the biological productivity of a resource cultivated by people (crops, forests) or exploited by people (fish), or by changing the physical life of a material used by people (such as steel, zinc, and marble), or by changing the enjoyment people derive from an environmental resource. Policies that reduce the level of acid deposition could yield benefits or in some cases costs by reducing the effects that deposition has on the biological productivity of these resources or on the physical life of materials.

Two general analytical approaches have been developed in past research to measure the dollar value of environmental impacts such as these: the market approach and the nonmarket approach. In the market approach, the value of an environmental change is inferred directly from observable market data on quantities and prices of particular goods. To measure the value of an improved crop yield associated with reduced acid deposition, for example, the policy analyst could estimate a model of demand and supply for the particular crop in which the crop yield affected supply. This model could be used to calculate the changes in consumers' and producers' surplus that are caused by yield changes of specific magnitudes. This information in conjunction with information about changes in crop yield associated with a specific policy, would be used to forecast the benefits of a particular deposition reduction policy on crops.

The market approach is most useful in valuing environmental changes when the environmental commodity is an input to the production of a commodity that is traded in an organized market with observable prices and quantities. This condition is satisfied for agriculture and commercial forestry. In each case, deposition is an input to the production process, and the output from this process is sold in organized markets. Econometric or other types of models of specific crop and timber markets can be developed to forecast consumers' and producers' surplus in the corresponding market. Crop and timber yields typically enter such models as an explanatory variable in the supply equation. The yield variable provides the link between the economics and the physical/biological effects component of the acid deposition program.

The market approach is also useful in valuing environmental changes when the affected environmental commodity directly or indirectly enters consumers' utility functions and consumers display their preferences for the commodity in related markets. Using the hedonic price technique, the value of mortality reductions associated with sulfate concentration reductions, for example, might be inferred from observable data on wages and salaries and from the probabilities of accidental death in various industries or occupations.

The nonmarket approach has been developed during the last 30 years by economists to value changes in the provision of environmental commodities that cannot be valued using market techniques, i.e., to consider commodities whose value to society is not reflected in market prices or quantities of related commodities. The approach relies heavily on direct surveys of consumers. These surveys obtain data from which a demand function for the affected environmental commodity can be econometrically estimated, or respondents are asked directly what they are willing to pay for an environmental change of a particular magnitude. One example of the first survey approach is the travel-cost method. This method is used frequently to assign a value to environmental changes that cause changes in the attractiveness of a particular recreational site or set of sites, for example, to value a change in the number of fish in a lake that might be caught by fishermen. Respondents are asked how many times they visit the recreational site in question and the costs they incur to travel to it. Information about site characteristics is also collected. Equations relating the number of trips to a site to the cost of traveling to the site and the site characteristics, i.e. demand functions, can be estimated using this information. These equations can then be used to calculate the aggregate consumer surplus with specific levels of the site characteristics and thus the change in consumer surplus, in one or more site characteristics.

The second survey approach is the contingent valuation method. A hypothetical market for an unpriced environmental commodity is constructed and respondents are asked how much they are willing to pay to "consume" specific quantities of the commodity. Willingness to accept compensation is also occasionally requested. For example, respondents may be asked how much they are willing to pay to increase the visual range in their hometown to 10 miles, to 15 miles and so on. Respondents' bids can then be statistically analyzed to estimate the benefits of such environmental improvements.

COSTS AND IMPACTS OF EMISSIONS REDUCTION STRATEGIES

NAPAP is currently funding development and refinement of several models that consider the costs and other impacts of emissions reduction strategies. These models will be able to forecast emissions of the major acid deposition precursors and emission control costs by pollutant, state, and emitting sector; they will also be able to predict these emissions and control costs under alternative policy scenarios. In addition, these models will be able to predict the changes in production and employment in affected industries associated with specific emission reduction strategies.

Electric Utility Costs and Emissions

NAPAP is planning to use the Advanced Utility Simulation Model (AUSM) to forecast baseline emissions of deposition precursors for electric utility plants for the 1985 Assessment, and to predict the plant-level changes in annual emissions and the corresponding capital and operating costs associated with specific reduction strategies in the 1987 report to Congress. AUSM will also be used in the 1987 Assessment to predict policy-induced changes in electric utility prices, production, and employment by state, as well as changes in state-level coal production.

Industrial Sector Costs and Emissions

For emissions and control costs related to industrial combustion processes, NAPAP is planning to use the Industrial Combustion Emissions Model (ICE) to produce information similar to that forecast by AUSM. ICE currently considers only emissions of sulfur and nitrogen oxides from industrial boilers. Emissions and costs from industrial process heaters will be added to the model. Other models are also being developed as part of NAPAP to consider other industrial emissions of SO_x and NO_x as well as industrial emissions of volatile organic compounds.

Both ICE and AUSM, as well as those models currently under development, can consider a fairly wide range of emission control strategies. Both are based on detailed inventories of existing capital stocks and current emissions. Both assume that future generating or productive capacity is added in a least-cost manner to satisfy increased demand, and both assume that emission regulations will be satisfied by individual plants or industries in a least-cost manner. Finally, these models can be run parametrically to obtain cost-emission reduction curves for each electric plant or industry by state. These curves can be used to find the least-cost way to allocate a desired reduction in national emissions to states and to allocate a state reduction target to individual plants and industries. The models consider a number of methods to reduce emissions, including fuel switching, installation of emissions control equipment at existing plants, retirement of existing plants with replacement by new plants, and other methods such as coal washing prior to combustion.

NAPAP is funding development of models to predict emissions of sulfur and nitrogen oxides and volatile organic compounds and related emission control costs in the transportation sector and residential and commercial sectors. Both models will forecast at the state level. Because few proposed policies have targetted emissions in these sectors, these models emphasize forecasting baseline emissions in order to forecast baseline deposition fields. However, the models will be able to consider the costs of emission reductions in the corresponding sectors.

Sector Integration

NAPAP is currently funding development of the Emissions Strategy Integration Model (ESIM), which will integrate the emission-cost models of the individual sectors. ESIM can accept from the sector models estimates of the cost associated with numerous specific emission reductions for each state and sector. It also can accept user specified deposition or emission targets by receptor state and deposition type and determine the level of emissions that 1) satisfies the targets and 2) minimizes the control costs for each state and sector.

Secondary Impacts

As discussed above, AUSM predicts state-level, policy-induced changes in electricity prices, production, and employment and in coal production. Because changes in state- and substate-level employment and personal income are carefully considered in the policymaking process, NAPAP is currently funding development of the capability to predict changes in these two variables (by state and industry) caused by specific policies. Specifically, NAPAP is funding expansion of the Metropolitan and State Economic Regions (MASTER) Model so that it can estimate changes in coal industry employment and earnings, as well as changes in employment and earnings in other sectors that 1) are sensitive to electricity price changes or 2) are sensitive to the changes in personal income caused by coal and electric utility employment changes. MASTER will be capable of generating such secondary impact estimates for each of the 312 Standard Metropolitan Statistical Areas (SMSAs) and 48 non-SMSA Rest-of-State Areas (ROSAs) in the contiguous United States, as well as for each of the 48 contiguous states. In addition, MASTER is being expanded to consider the impacts on employment and income associated with changes in production of emission control equipment, crops, and timber.

BENEFITS OF EMISSION REDUCTIONS

Since information on economic consequences is specifically requested in NAPAP's originating legislation, and NAPAP believes information on the economic benefits of reducing deposition is an important piece of policy information, the program is sponsoring a number of projects to estimate the value of the benefits of reduced acid deposition. Most of this research is designed to measure benefits rather than effects. Several NAPAP Task Groups are responsible for the effects research; the economics program is responsible for ensuring that the effects information results are presented in a form usable by economists for the actual valuation. A less than fully quantitative approach could be necessary if the physical effects inputs cannot be delivered.

NAPAP's economics program has focused on four areas: aquatics, materials, forestry, and crops.

Aquatics

Numerous studies have indicated it is likely that acid deposition lowers the pH of affected rivers and lakes. The actual dose-response relationships between the pH of the lake or stream and the health of the aquatic biology is less clear. However, available evidence suggests that acid deposition has reduced the number of fish and other aquatic life in lakes susceptible to acid deposition. If this hypothesis is valid, we would expect that a reduction in fishing opportunities is one of the impacts of acid deposition.

Potential Techniques. Techniques for valuing reduced fishing opportunities include several varieties of travel-cost methods, a participation model linked to a valuation model (The Russell-Vaughn Approach) and contingent valuation approaches. Each of these techniques has particular strengths and weaknesses that lend themselves to particular research questions.

The travel-cost technique, usually attributed to Clawson-Knetsch, attempts to infer the shape of the demand curve by examining data on how people respond to different travel costs. This method has usually been applied to valuation of changes in fishing at specific sites. The regional effects of acid deposition mean that a disaggregated site-by-site approach is not useful. Other techniques, including characteristic approaches such as Morey's (1981) or models developed by Samples and Bishop (1983), display some promise for assessing at least large geographical effects.

Alternatively, both the Russell-Vaughn and contingent valuation approaches seem more appropriate for large regional or national assessments. The Russell-Vaughn model is a participation model. In general, a participation model links an individual's decision to fish at a site with the site's characteristics. The Russell-Vaughn model specifically relates changes in fishing-days or some other measure of fishing participation to a set of dependent variables that include characteristics such as fishable acres or success rate. The participation equation can be linked to a unit value for fishing days to obtain estimates of the value of changes in fishing opportunities. Successful use of the model requires knowing how acid deposition affects one or more of the dependent values, for example how fishable acreage affects fishing days. The third potential technique for valuing recreational fishing is a contingent valuation or "bidding game" model, discussed at greater length in the section on ecosystem values.

The 1985 Assessment. The 1985 Assessment will focus on those geographic areas where research indicates fish populations and recreational fishing have the greatest risk of being affected. This is the intersection of three sets of data, data on

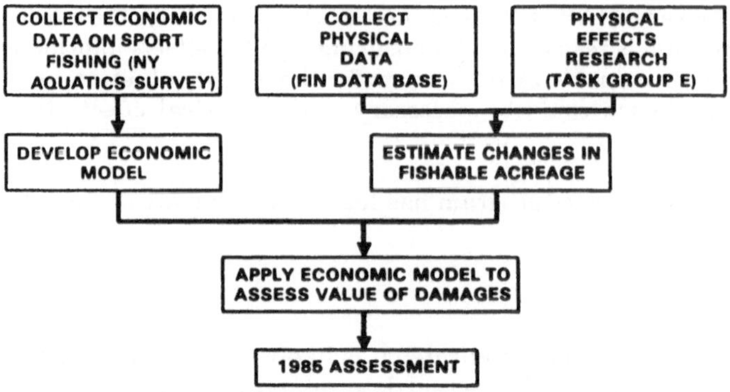

Figure 1. Framework for Economic Valuation of Recreational Fishing

acid deposition, sensitive surface waters, and sensitive fish species. Currently the area slated for this type of work is the high altitude Adirondack lakes region.

A current NAPAP project is producing a statistical travel-cost model based on Adirondack historical data, which relates a measure of economic value, such as a consumer surplus measure, to the characteristics of each fishing site. If successful, the model will assess the economic damages to recreational fishing from acid deposition expressed in dollars. To perform the calculation the model would need data on site characteristics such as the total fishable acreage that would be available in the absence of any acid deposition from man-made sources. This effects data will be obtained through interactions with the task group working on physical effects (see Figure 1).

An alternative or perhaps an addition to the travel-cost model is the use of a model linking acid deposition and changes in participation. In this approach, changing the level of fishable acres changes the number of fishing days. The estimates of changes in participation can be linked to an imputed unit value for fishing days from other studies to obtain an estimate of regional damages. Potential sources for imputed values include contingent valuation surveys on the value of fishing opportunities and other studies of fishing values, such as travel-cost studies. A decision on the final methodology, participation model or the travel-cost model, will be made based on a review of the preliminary results of work underway. Either technique should enable us to have estimates of economic damages for the 1985 Assessment. If the physical effects information on the changes in fishable acres is not available, we plan to estimate the monetary damages for a range of possible effects. For example, we might estimate the monetary damages of 2 percent, 5 percent, 10 percent and 20 percent changes in fishable acres.

The 1987 Assessment. Since the 1987 Assessment is to be a national assessment, the Russell-Vaughn approach or a contingent market study will be more appropriate than the travel-cost method for that document. One potential data source is the 1980 U.S. Fish and Wildlife Survey of Fishing, Hunting and Wildlife Associated Recreation. This data source, which is national in scope, has been used by Russell-Vaughn and other economists in valuing recreational damages. Work on developing a final approach will begin in the next few months.

Materials

In the area of materials, NAPAP is attempting to integrate three elements:

- dose-response information linking material loss to an agent or element in the environment;
- inventories of materials susceptible to damage, including location, specific materials, and structural function; and
- economic methods to convert estimates of physical damages into economic terms.

The program distinguishes between common construction materials and cultural buildings and structures. Each of these requires different types of economic valuation methods.

Potential Techniques. Three techniques have been reviewed for their usefulness in NAPAP assessments for common materials: the damage function approach, the optimal materials use approach, and the commercial property value approach. Techniques for valuing damages to cultural structures are treated separately.

Damage Function Approach. The damage function approach uses information on physical damages in a straightforward way to arrive at damages expressed in economic terms (see Figure 2). Several of the steps in Figure 2 are the responsibility of physical scientists; for example, estimating damage functions is performed by the task group studying materials. The economic component needs to identify repair, replacement and maintenance costs expected to change as a result of the physical damage, estimate these costs, and aggregate the physical effects and cost information. One problem with the damage function approach is that it is limited to materials for which physical damage functions are available. Many functions simply do not exist. Also, this approach does not consider any responses by owners to the increased damages. Other alternative approaches that do not require a direct measure of physical effects are available to estimate the economic damages, including a property value approach. This approach, discussed later, uses secondary data to derive associations between pollution and market effects.

The Optimal Materials Use Approach. The "optimal materials use model" is a generalized form of the damage function approach. The generalization occurs in two areas. First, the approach explicitly recognizes that there are two levels of decisions about materials: decisions to use different materials, and decisions on how to maintain a given material once it has been chosen. The optimal materials use model can address both sets of decisions. Furthermore, the model incorporates the possibility that air quality can influence decisions at either stage. That is, air quality can influence the choice of a particular material (e.g., aluminum siding vs. painted wood) as well as the maintenance practice for the material (e.g., frequency of repainting). Thus, the model emphasizes that measurement of the full economic effects of air pollution may require analysis of both stages.

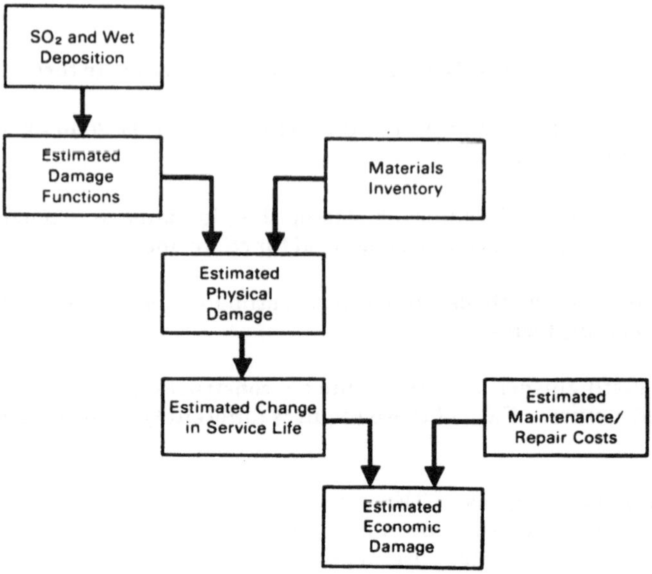

Figure 2. Framework for Economic Assessment of Material Damages

The second generalization occurs in the analysis of the maintenance decisions. This analysis recognizes that decisions about maintenance should, in principle, consider such things as the fact that (1) more frequent maintenance may keep the material closer to original performance levels; (2) more frequent maintenance may extend the useful economic life of a material; (3) less frequent maintenance will delay incurring maintenance cost; and (4) less frequent maintenance may increase the costs of maintenance when it is performed and in the extreme case, it may require total replacement of the material.

Unfortunately, the generalization and increased sophistication require large increases in information to perform the analysis. Additional required information includes how pollution affects the choice of materials or the maintenance of the material.

The Property Value Approach. The property value approach, using multiple regression, is a commonly used technique to estimate the effects of acid deposition on the residential sector. The approach assumes that geographical variations in property values can be explained by attributes such as location, age, condition and air quality. A major advantage of this approach is that an explicit damage function is not required. The property value is regressed against values of the explanatory variables, including one for air quality. The coefficient on air quality is interpreted as the value of a small change in air quality as reflected on property values.

The technique has not been used to estimate the effect on commercial or non-residential buildings because of a lack of data. Non-residential property differs substantially in uses and the attributes that determine market values. It might be possible to limit the analysis to a more homogeneous group of buildings, for example, a certain type of commercial buildings. However, other techniques would be required to perform a comprehensive analysis that included other building types.

The property value approach has several other problems. For air pollution studies, visibility, health effects and material effects would probably be included in the coefficient on air quality. For example, housing prices may be lower in a heavily polluted area both because the pollution damages the building material and because consumers might perceive health effects or visibility degradation. The analysis would need to avoid adding visibility or health as a separate effect. Also, since many common structures such as bridges and public buildings are never bought or sold, the technique is limited to buildings with available market prices.

Approaches for Cultural Structures. The economic assessment of cultural damages is complicated by the inability of the damage function approach to fully capture an individual's willingness-to-pay to avoid damages to certain buildings or structures. These buildings or structures have what economists call "option" or "existence" values. For example, in spite of never using a good, an individual might be willing to pay to preserve the option of future use of the good or to preserve the good for the use of future generations. Some outdoor art objects, gravestones or monuments may have significant intrinsic values. For example, an individual may place a high value on preserving the writing on a relative's gravestone for future grandchildren. If the maintenance is not sufficient to prevent degradation, the value of the lost gravestone is significantly more than a damage function would indicate.

Estimating the intrinsic values will require a contingent valuation study or other techniques to obtain estimates of individual's willingness-to-pay to preserve cultural structures. Contingent valuation or "bidding game" studies attempt to obtain estimates of individual's willingness-to-pay through surveys and direct questioning rather than through examining how people respond in markets. While such surveys need to be carefully prepared and written to avoid biases, the technique has been used with various levels of success to obtain estimates of willingness-to-pay.

The 1985 Assessment. Following a review of the possible approaches discussed above, NAPAP decided to use the damage function approach for the 1985 Assessment. It was chosen because it easily integrated and closely linked to the physical damage research. Although it is not fully consistent with the "willingness-to-pay" norm for estimating the benefits of reduced pollution, more complex models could not be implemented on a regional scale.

The damage function method to be used to value damage to common construction materials is depicted in Figure 2. It shows one of the two ways the physical science and economics data can be processed. The other way is discussed below.

The economic analysis seeks to take advantage of the insights of optimal materials use models and life-cycle cost models. Essentially, these generalize the damage function method, focusing on damage as it affects the timing and choice of repair, maintenance, and replacement decisions. These approaches will be used to develop estimates of economic damages in the northeastern U.S. using the flow of information illustrated in Figure 2. A sensitivity analysis will be performed at the final stage. This will be useful for developing additional information on error bounds and determining areas where further improvements can be made in future assessments.

An estimate of the economic damages to cultural buildings and structures will also be included in the 1985 Assessment. The scope of this effort will be limited geographically, by material or structure at risk, and by economic data that do not account for nonowner, nonuser intrinsic values. The results will utilize estimated costs of repairing and maintaining cultural structure at risk to acid deposition. The damage function sequence depicted in Figure 2 will be the basis for valuation, with the difference that increased repair frequency is the result of an aesthetic change rather than a change in service life.

Since the translation of material loss on cultural structures to changes in maintenance or repair intervals depends on aesthetic considerations, this part of the damage function linkage is less certain than the corresponding task for building materials. Current research on cultural effects is focused on estimating damage functions, with the task of ascertaining intrinsic values postponed for future assessments. In any case, the cost of maintenance and repair data can give some indication of the magnitude of damages, even without precisely accounting for repair frequency, and this cost data will form the basis of the limited assessment of cultural resources for the 1985 Assessment.

The 1987 Assessment. For economic valuation of material damages, in 1987 and subsequent Assessments, NAPAP will build on the approach used for the 1985 Assessment. Improvements will include better representation of the differences in repair and maintenance schedules due to acid deposition, increased materials, increased material uses, increased linkages back to the physical effects research, and increased building and material inventory information. At present, several computer linkages are being investigated that would allow fairly rapid evaluation of the effects of various pollutant levels on material damages and the economic valuation of those damages.

For cultural structures, the important advances will likely include attempts at estimating intrinsic values associated with cultural structures not receiving repair or maintenance. Estimating the intrinsic values would likely require a contingent valuation study. NAPAP has a project underway to investigate the numerous issues associated with conducting a contingent valuation study of acid deposition damages.

Forestry

Several studies have noted a decline in the tree ring growth rates in the eastern U.S., and forest dieback at certain high elevations (OTA 1984, p. 224-235). In Germany and other European countries the problem is even more severe with major evidence of visual damage. The exact cause of these changes has not yet been determined, but many hypotheses point to pollution, if not acid deposition, as the cause. The following paragraphs discuss NAPAP efforts to value the damages of forest growth slowdowns.

Potential Techniques. Changes in forest productivity is of direct consequence to society because of the loss of products derived from wood. Market models that attempt to replicate the market process are generally considered the proper way to estimate the value of environmentally induced changes (Freeman 1979, p. 255). We first review a more generic breakdown of market models into normative and positive and briefly discuss the advantages and disadvantages of each for our task.

Generally, economic models are distinguished by the following characteristics: (1) whether they are dynamic or static; (2) whether they are designed primarily for policy analysis or for forecast of short-term movements in prices and output; 3) whether linkages to the national economy in the model are explicit or implicit; and 4) whether the geographic delineation contained in the model is national or regional. These distinctions become important when selecting a model to fulfill a specific set of research objectives. However, because an individual model may incorporate a great many different attributes, classifying timber market models along multidimensional lines often obscures more general differences. Therefore, we have settled on distinguishing between models constructed so as to show how economic agents *should* behave in order to achieve a given set of economic and/or non-economic objectives and those designed to predict the *actual* behavior of economic agents within timber markets. The first type of modeling follows the normative approach and the second is the so-called positive approach. These positive models usually involve econometric models of market behavior.

The 1985 Assessment. Our review of normative models indicates several features which make it difficult to consider their use in a national assessment of the effects of acid deposition on timber product and stumpage markets. Fortunately, at least one positive model is available that attempts to model the supply and demand side of timber products and timber stumpage. This is the Timber Assessment Market Model (TAMM), which was developed for the U.S. Forest Service to provide long-range projections of U.S. timber inventories, timber products output, and prices. The specific attributes that make TAMM unique in the application under discussion are as follows:

- Multi-Regional Structure. TAMM is a national model, but it is broken down geographically by product regions.

- Multi-Market Structure. TAMM contains explicit submodels for regional stumpage markets and regional product markets for both hardwoods and softwoods.

- Multi-Market Spatial Equilibrium Features. In TAMM, the determination of market clearing prices and output levels in different markets in different regions allows for physical flows of harvested timber from stumpage markets to product markets and for flows of lumber and plywood from product supply to product demand regions.

- Direct Linkage to Inventory Projection Model. TAMM is directly tied to an inventory projection model, the Timber Resource Assessment System (TRAS). Timber removals in TAMM depend on inventory levels shown by the inventory model; removals made by TAMM are then fed back into TRAS to update the inventory for the subsequent period.

- Dynamic Properties. Finally, TAMM is an annual model; however, it is also a dynamic model in the sense that some of the variables determined by the model in the current period (t) are used as exogenous variables (e.g., as model inputs) in the subsequent period (t+1). Thus, the model can simulate market behavior over time.

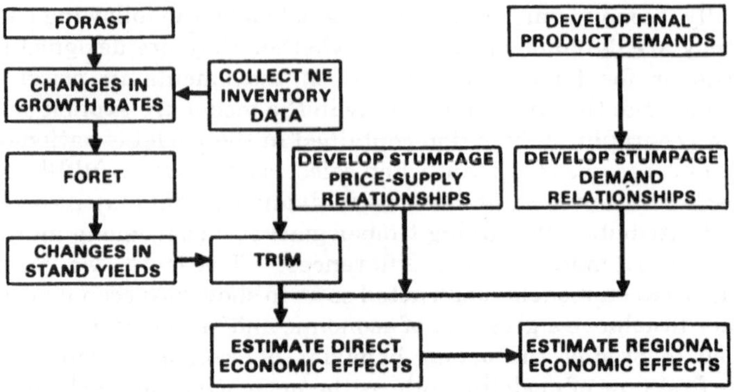

Figure 3. Framework for the Economic Assessment of Forestry Damages

We investigated other models that possessed some of these features, but none of them combine all of the features believed necessary to conduct an assessment of the effects of acid deposition on regional stumpage and product markets in the United States. Because TAMM was designed specifically for conducting assessments similar to the ones planned for acid deposition, this advantage is not surprising.

The 1985 Assessment will be limited to the northeastern and southeastern United States. The forest inventory information has been obtained from the U.S. Forest Service and work is underway linking TAMM with the forest inventory model, TRIM. The outputs from TAMM include regional price and quantity information, consumer and producer surpluses and harvest information. Figure 3 shows the forestry economics portion of the 1985 Assessment. Again, the economics tasks require inputs from research on the physical damages to trees due to acid deposition. Should this research (FORAST) not prove conclusive or be able to distinguish acid deposition damages from other damages, the assessment will be less quantitative.

1987 and Subsequent Assessments. The national assessments will continue to use TAMM in conjunction with an inventory model. Additional regions of the country will be added to the Northeastern forest inventory collected for the 1985 Assessment. The subsequent assessments will also try to evaluate the effect of decreased growth rates of the magnitudes induced by acid deposition on individual forest owners' harvesting decisions. The European experience is that forest harvest accelerates at the onset of decline in the forest's growth rate. This behavior could have important implications for southeastern U.S. plantation forests.

Crops

Research on whether acid deposition reduces crop yields has produced mixed results. Some crops appear to experience yield losses, others no change, and still others experience yield gains when exposed to rainfall with a low pH. The consistent conclusion is that, under ambient conditions, any yield loss is relatively low, below the yield reductions caused by ozone and other pollutants and certainly below the damage caused by pests and unusual weather conditions. However, even small yield losses could translate into large dollars because of the total value of U.S. agriculture.

Potential Techniques. A number of existing agricultural models are capable of capturing supply adjustments and corresponding price and quantity effects in the agricultural sector. The mathematical programming and econometric approaches have been widely used in agricultural and environmental economic analysis. Existing models using math programming are able to capture price and substitution effects and provide estimates of distribution. Similarly, econometric models either exist or could be easily modified to evaluate the consequences of changes in crop yields. These models construct the supply and demand functions using econometric techniques on existing data.

An alternative is the hedonic approach, which generally requires less dose-response information than either the econometric or math programming approaches. The hedonic approach is commonly used to measure the effects of pollution on urban property markets. The process is similar for examining the relationship between pollution and rural or farm values. The market price of a number of farms would be regressed against the attributes of the individual farms, including one reflecting the level of pollution. We know of no study that has used the property value approach for valuing pollution effects on cropland values. Controlling for the numerous factors affecting crop productivity and the other variables affecting a farm's market value would require large amounts of data in order to detect the expected small change in yields due to air pollution. Also, the hedonic approach does not use scientific data on the dose-response to link the physical effect research to changes in consumer and producer surpluses. Some feel this limits confidence in the regression coefficients.

The 1985 Assessment. Acid deposition damages expressed in dollars could be estimated at various levels of sophistication. For the 1985 Assessment, we will use a combination of micro- and macro-economic models that will allow us to capture the responses of farmers to price changes and link those changes to a national model of crop supply and demand.

We are using this approach for several reasons. While the less sophisticated models, for example multiplying market price times yield changes, can be run with easily obtained data, they have two drawbacks important for assessing the economic effects of acid deposition on crops. First, the simple models assume that any supply changes resulting from acid deposition do not affect price. Since the demand curves for crops have a low price elasticity, that is, are relatively insensitive to price, a small percentage change in supply, such as believed to occur with acid deposition, results in a larger effect on prices. Second, they are based on incomplete and inadequate methods of representing changes in welfare due to acid deposition on crops. The simple models fail to include any consumer effects, to consider producer responses to changing prices or yields, or to adequately estimate effects on different producing regions. A complete analysis will require estimating the effects on consumers since these responses will be of the same magnitude as the effects on producers. On the other hand, producer responses may be ignored if they prove small in comparison to the overall effects, although they cannot be ruled out without some research into the magnitude of the potential response. Finally, we are using this approach because relatively complete agricultural models that have been developed for other purposes can be used with little modification; thus, the costs of modifying and using more sophisticated models are not overwhelming.

The economic assessment will focus on the direct effects in the southern producing states. The assessment will be limited to soybeans. An objective of ongoing economic research is to develop the capability of performing other assessments at the national level or for other regions. These regions account for the vast majority of the U.S. production of soybeans and could be affected by acid deposition. Models developed for the National Crop Loss Assessment Network (NCLAN) will be used to obtain estimates of consumer and producer surplus (Adams, Hamilton, and McCarl 1984). Several dose-response functions will be used to provide a sensitivity analysis.

The NCLAN farm sector model simulates farmer behavior based on assumptions of profit maximization. The model under consideration for this activity is REPFARM, an existing farm level model (McCarl 1982). Environmental effects can be included through changes in yields in the REPFARM. This model will be used to develop supply responses at the regional level as inputs to a more aggregated, macro model. The NCLAN macro model, with several adaptations, will be used to trace total effects on consumers and producers. The outputs will be an estimate of total changes in consumer and producer surplus and an allocation of effects among producer groups. Figure 4 shows the framework for the crop economics portion of the 1985 Assessment.

The 1987 Assessment. Because the most recent experiments on the changes in crop yields due to acid deposition have revealed at worst a very small reduction, the economic analysis for crops is being discontinued following the 1985 Assessment. NAPAP is also reducing funding for physical effects work in this area. Research would resume if continuing experiments should show larger crop effects than present experiments indicate.

Other Ecosystems

Acid deposition reduction policies yield a number of benefits not easily categorized into the NAPAP categories of crops, forests, materials, and recreational

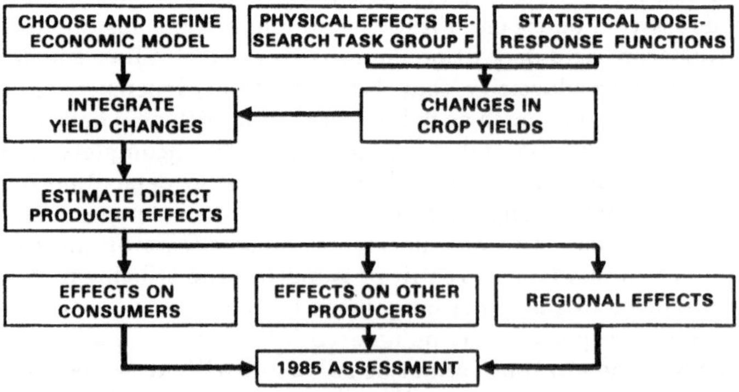

Figure 4. Framework for the Economic Assessment of Crop Damages

fishing. Two such benefits come to mind. First, if deposition reductions increase the foliage as well as the population and health of animal and plant species, then the aesthetic enjoyment that campers, sight-seers, bird-watchers, hunters, and others derive from specific recreational activities are also increased. In other words, these policies may have benefits associated with current and future use of recreational resources in addition to the recreational fishing benefits discussed above. Second, changes in the quantity or quality of these resources may have intrinsic value. An individual may be willing to pay some amount to increase or improve these resources even though he or she does not currently use the resource and does not expect to use it in the future. Intrinsic values of this type arise from two sources: (1) individuals may be willing to pay to preserve the option to use the resource in its current or improved state in the future; (2) individuals may be willing to pay to preserve the existence of the resource in its current or improved state.

Techniques. The benefits of reduced damages to other ecosystems can be estimated using either the travel-cost method or the contingent valuation method, the two nonmarket valuation techniques. For the travel-cost method, specific recreation and aesthetic resources affected by acid deposition and specific deposition reduction policies would be identified. A travel-cost demand model would be estimated for each and the resulting model used to forecast the use-related consumer surplus associated with improvements to the site. In the contingent valuation method, visitors to a particular recreational or aesthetic area could be asked how much they would be willing to pay to have the area improved in a particular manner by a specific magnitude. These bids could be statistically analyzed to develop a model that could be used to predict the user-selected benefits of specific improvements to the area.

NAPAP currently plans to use the contingent valuation method to estimate these user benefits. The travel-cost method has not been used to estimate this type of benefit, primarily because the "sites" are not typically distinct, as in the case of a lake or a specific park, but rather might include the entire eastern U.S. A large number of "sites" would have to be considered to handle a large regional analysis. The contingent valuation method appears to be able to handle these two problems more easily than the travel-cost method and it has been used by economists in the past to measure intrinsic values. The intrinsic value of improving a particular site can be estimated by surveying both users and non-users of the resource at a location, such as a major city or set of cities near but not directly adjacent to the site. As in the case of user benefits, bids from this survey can be statistically analyzed to develop a model that can in turn be used to predict the intrinsic value of a specific resource enhancement.

1985 and 1987 Assessments. Quantitative estimates of monetary damages will not be generated for the 1985 Assessment. However, the assessment will include a discussion of these effects. Current plans are to include both user and intrinsic benefits of other ecosystems in the 1987 Assessment. In support of this, NAPAP is currently funding a multiyear contingent valuation research project. During FY 1985, important issues to be considered in a contingent valuation study of other ecosystem benefits will be identified. Two important issues are being studied: (1) the bias associated with contingent valuation estimates, and (2) how best to include option and existence values in the benefit estimates without double counting. The latter issue arises because such values are typically associated with prevention of irreversible environmental impacts, which may or may not be relevant to the acid deposition phenomenon.

During FY 1986, the present research plan is that the survey instrument will be applied and a set of statistical models explaining individuals' bids will be developed. Samples of the populations of several northeastern states will be surveyed; respondents will be asked to state their willingness to pay for specified improvements to several relatively large, important, recreational/aesthetic areas, such as Adirondack Park. Care will be given in distinguishing between use and intrinsic values. A statistical model that explains and predicts the individuals' bids for improvements to each area will be developed. During FY 1987, these statistical models will be used to estimate, in monetary terms, the benefits associated with specific policies to the areas considered. The efficacy of extrapolating benefit estimates for these areas to the nation as a whole will be investigated. These plans are currently under review and may be scaled down for budgetary reasons.

Ancillary Benefits

In addition to reducing the damages to forests, crops, and aquatic systems that are directly caused by acid deposition, proposed emission reduction policies may have physical, biological, and economic impacts that are not a direct result of acid deposition reductions. These impacts may be of significant value to society and thus should be considered in assessments of the corresponding policies.

Definition and Examples. Ancillary benefits of deposition reduction policies are those impacts, benefits or costs, which are not directly related to acid deposition, and are of economic value to society. Examples include improved visibility, improved crop and timber yields apart from those caused directly by acid deposition, and reductions in human mortality and morbidity. Each of these phenomena may result from the reduced ambient sulfate concentrations that will likely accompany policy-induced reductions in sulfur dioxide emissions. Unfortunately, there is not a consensus on the methodology for evaluating these effects. The value of improved visibility, decreased crop, timber, or material damages, and the effect of sulfur dioxide emissions on human mortality and morbidity are not firmly established. However, to provide an estimate of the possible magnitude of the effects, we have selected estimates from the works of Latimer (1984), Tolley (1983) and Freeman (1982) as the basis for calculation. These are some of the most recent works that address these issues in a quantitative fashion.

Recent studies of air pollution control policies suggest that the ancillary benefits of proposed acid deposition reduction policies may be quite large. Latimer et al. (1984) recently calculated the sulfate concentration reductions by state for a 31 state, 12-million ton reduction in sulfur dioxide emissions. These sulfate concentration reductions would be expected to increase visibility by about 16 percent across the 31 states (Latimer 1984, p. 7-58). Using the results of a recent study by Tolley et al. (1983), Latimer et al. (1984) then estimated that the value of such a visibility improvement would be $1.9 billion annually. Freeman (1982) estimated that a reduction of 13,900 deaths per year[1] would result from sulfate concentrate

[1]The Freeman estimates are based on an earlier study by Lave and Seskin (1977). The Lave and Seskin work has been very controversial with little scientific concensus regarding their estimates.

reductions which were less than those used by Latimer et al. (1984). Using $1.59 million as the value of a statistical life (Freeman 1982, updated to 1984 dollars), the value of these mortality reductions is $22.1 billion annually. These works suggest that the ancillary benefits of a 12-million ton sulfur dioxide emission reduction may be in the neighborhood of $24 billion per year. Should these estimates be high by an order of magnitude, the ancillary benefits from acid deposition reductions are still large. Clearly, additional work is needed to provide a more definitive basis for evaluating the ancillary benefits of reduced sulfur dioxide emissions.

1985/1987 Assessments. While NAPAP will not provide estimates of ancillary benefits for 1985, current plans are to include ancillary benefits in the 1987 Assessment in a limited manner. There are no plans to include health-related ancillary benefits because of the controversy that surrounds the valuation-of-life issue and, particularly, because of the perceived weak state of the science in estimating changes in mortality and morbidity caused by ambient air quality changes. At this time, NAPAP plans to limit consideration of ancillary benefits only to those associated with sulfate concentration changes because the atmospheric dispersion models being developed as part of NAPAP are primarily sulfur oxide models.

Thus, the only ancillary impacts to be considered in the 1987 Assessment will be changes in visibility, crop yields, and timber yields associated with sulfate concentration reductions. These benefits will be estimated in a series of steps. First, it is planned that the NAPAP atmospheric models will be used to predict average annual and annual peak sulfate concentrations by region, both for the base case (i.e., no new policy) and various policy-related emission scenarios. Second, these sulfate concentration changes will be used to estimate annual changes in visibility, crop yields, and timber yields by region (and crop/timber species). Third, the dollar value of these changes will be estimated. For crop and timber yield changes, the economic models developed for NAPAP to value deposition-induced damages will be utilized. For visibility, preliminary plans call for modifying or developing a model that can predict the willingness-to-pay of individuals for visibility improvements of specific magnitudes. This will be used to provide information on the value of visibility improvements in the 1987 Assessment.

CONCLUSIONS

We believe there are six main points or conclusions about NAPAP and the role of economics in it that deserve attention.

Integrated Assessment

One of the objectives of Task Group I is to perform integrated assessments. Economists are attempting to value only those impacts or deposition damages that physical and natural scientists in NAPAP are analyzing; for example, because health effects have received little study within the program, research on valuing health effects has not taken place. On the other hand, economists are attempting to value all of the effects that physical and natural scientists are studying. In these areas, interactions between economists and the physical scientists are crucial for good integrated assessment and research planning. If the scientists studying the effects of acid rain are unable to identify precise damaging effects, the economics portion of the assessment will be based on economic models, when appropriate, which will estimate the monetary damages of several levels of physical damages. This will provide information on the range of damages and clues regarding where to place future research efforts.

Objectives of NAPAP Economic Research

Because NAPAP was organized as a research and assessment program, the role of economics in NAPAP is as a provider of information to decisionmakers, and the economic research performed in NAPAP clearly reflects this role. Models are not developed unless they help answer assessment questions and existing models are not available. The economic research is planned toward the 1985 and subsequent assessments. First, research is being performed to modify or develop models for each of the effects areas considered in NAPAP by physical and natural scientists. Thus, the economics program is exhaustive. Second, these models are capable of generating benefit estimates under alternative policy scenarios. Third, these models will be able to forecast benefits. Fourth, the models will be able to consider benefits to all regions of the country. Fifth, they are, to the extent possible based on methods and data that are consistent with economic and statistical theory and are accepted by both experts and laymen; the models are not experimental research efforts. Sixth, the models are being designed to consider uncertainty.

Cost Valuation

Techniques to value the costs of implementing emission reduction policies are well developed and well accepted. NAPAP is currently funding major research efforts to enhance existing models and develop new models that apply these techniques to acid deposition-related emissions.

Valuation of Market-Related Benefits

Techniques to value benefits related to crops, forests, and materials are fairly well developed and well accepted. Benefits in these areas can be valued using market-based techniques. NAPAP is funding small projects to modify existing crop and timber market models to consider the acid deposition phenomenon. A major effort in the materials area is required to develop a materials-at-risk inventory, to which established valuation techniques will be applied.

Non-Market Benefit Valuation

Techniques for valuing non-market benefits are not well developed nor well accepted. In particular, such techniques do not lend themselves particularly well to regional or national applications; travel-cost methods are more appropriate as small-region techniques. In addition, few models using these methods have been constructed that can forecast with data available from secondary sources. The contingent valuation method has rarely been used to develop predictive models; typically, sample mean bids are calculated, but it is not clear whether these means should be extrapolated to larger populations. Because of these problems, a rather large share of NAPAP's funding on benefit methods is dedicated to development of models to forecast national policy benefits related to recreational fishing and other ecosystems.

Ancillary Benefits

Recent research on the visibility and health benefits of sulfur dioxide emission reductions suggest that the ancillary benefits of acid deposition reduction policies may be larger than the deposition damage reduction benefits. Because of this and because models to forecast the value of national visibility improvements require additional development, NAPAP is examining ways of including the value of visibility improvements in its economic model. However, at present, there are no plans to include ancillary health benefits in the 1987 assessment.

REFERENCES

Adams, R.M., Hamilton, S.A., and McCarl, B.A., 1984, *The Economic Effec ts of Ozone on Agriculture.* Final Report to USEPA Corvallis Laboratory.

Freeman, M.A., III, 1979, *The Benefits of Environmental Improvement.* Resources for the Future. Baltimore, Maryland.

Freeman, M.A., III, 1982, "Air and Water Pollution Control: A Benefit-Cost Assessment," Wiley, New York.

Integrated Task Force on Acid Precipitation, 1982 (ITFAP 1982a), "Operating Research Plan for National Acid Precipitation Assessment Program," Draft. Washington, D.C.

Integrated Task Force on Acid Precipitation, 1982 (ITFAP 1982b), "National Acid Precipitation Assessment Plan," Washington, D.C.

Latimer, G.A., Anderson, G., Banks, J., Hogo, H., Ireson, R., Morris, R., Pollack, A., Saxena, P., Rae, D., and Rowe, R., 1984, *Visibility and other Air Quality Benefits of Sulfur Dioxide Emission Controls in the Eastern United States (Draft).* SYSAPP-831243. Systems Application Inc. San Rafael, California.

Lave, L.B. and Seskin, E.P., 1977, *Air Pollution and Human Health.* John Hopkins University Press for Resources for the Future. Baltimore, Maryland. 368 pp.

McCarl, B.A., 1982, "Cropping Activities in Agricultural Sector Models: A Methodological Proposal," *American Journal of Agricultural Economics.* 64:768772.

Morey, E.R., 1981, "The Demand for Site-Specific Recreational Activities: A Characteristics Approach," *Journal of Environmental Economics and Management.* 8:345-371.

Office of Technology Assessment, 1984, *Acid Rain and Transported Air Pollution.* U.S. Government Printing Office. Washington, D.C.

Samples, K.C. and Bishop, R.C., 1983, "Estimating the Value of Variation in Angler's Success Rate: An Application of the Multiple Site Travel Cost Methods." Working Report.

Tolley, G.F., Randall, A., Bomquist, G., Fabian, R., Fishelson, G., Frankel, A., Hoehn, J., Crumm, R., and Mensah, E., 1983, *Establishing and Valuing the Effects of Improved Visibility in the Eastern United States.* Prepared for the U.S. Environmental Protection Agency, Office of Research and Development.

APPENDIX

The National Acid Precipitation Program

The Interagency Task Force on Acid Precipitation (ITFAP) was established by the Acid Precipitation Act of 1980 to plan and manage the National Acid Precipitation Assessment Program (NAPAP). Planning for the National Program began in October 1980, when the Task Force was first organized. In accordance with the Act, the Task Force drafted a National Acid Precipitation Assessment Plan, outlining the proposed research. After extensive public review, the final version was submitted to Congress in June 1982.

The Task Force oversees the planning and implementation of NAPAP. The 20 members include one high-level representative from each of the 12 federal agencies in the Program, the directors of Argonne, Brookhaven, Oak Ridge, and Pacific Northwest National Laboratories, and four presidential appointees. The Task Force is jointly chaired by the National Oceanic and Atmospheric Administration (NOAA), the Department of Agriculture (DOA) and the Environmental Protection Agency (EPA). Other participating Federal agencies include: the Departments of Interior (DOI), Health and Human Services (HSS), Commerce (DOC), Energy (DOE) and States (DOS); the National Aeronautics and Space Administration (NASA); the Council on Environmental Quality (CEQ); the National Science Foundation (NSF); and the Tennessee Valley Authority (TVA).

The role of the Task Force in planning the interagency budget for the National Program is an effective and unique aspect of the federal effort. By working together through the Task Force, the agencies have established a research program that addresses national needs while building on the research expertise of the individual agencies. This interagency planning process attempts to avoid unnecessary duplication and crucial omissions in the National Program. However, at times it is also bureaucratic and cumbersome.

The Task Force sets the research goals for the NAPAP, identifies the projects needed to meet these goals, and decides which agencies are best suited to conduct the necessary work. The result is a comprehensive program of projects, with each agency contributing to specific aspects of the overall national effort.

The Task Force has 10 Task Groups, one for each of the nine categories in the national program and one for international activities. These technical groups include program managers and experts from the participating federal agencies and national laboratories. They are responsible for detailed planning and work in their assigned areas. Figure 5 shows the Task Force's operating structure and the relationships among the various agencies and task groups.

All but one of the task groups shown in Figure 5 focus on the advancement of scientific information regarding the nature, causes, effects and controls of acid precipitation. Their activities center on gathering and analyzing data, studying the various physical and ecological processes of the phenomenon and developing adequate predictive models. Only the Assessments Task Group (Task Group I) is directed towards integrating the research results, focusing the research upon policy issues, and assessing the potential results of various measures to reduce acid deposition.

The goal of the Assessments Task Group research program is to build upon earlier studies and draw from existing analytical methods to construct the framework for comprehensive assessments. This capability will be helpful in at least two ways. First, it will be designed to deal quantitatively with the range of uncertainty around various data and their use. Research and assessment will continuously track, guide and inform each other as uncertainties decrease over time. In addition, developing methods to organize scientific results and applying them to policy questions early in the research program will ensure that the National Program can produce information relevant to making informed policy decisions.

Task Group I is responsible for several reports and assessments of increasing sophistication over the next 5 years. The first will be completed in 1985, followed by subsequent assessments in 1987 and 1989. Figures 6 and 7 show how information for 1985 and 1987 and subsequent assessments will be organized.

NAPAP has defined the scope of the major assessment in 1985 to encompass three areas: (a) an assessment of economic and physical damages attributed to acid deposition; (b) an analysis of uncertainties associated with emissions and atmospheric processes; and (c) a description of the framework for the integrated assessment methodology which will be the basis of the 1987 and 1989 integrated assessments. Because of the scientific uncertainty of linking emissions completely to effects, the 1985 Assessment will be modular; that is, each segment of the phenomenon will be described without attempting to integrate the analysis.

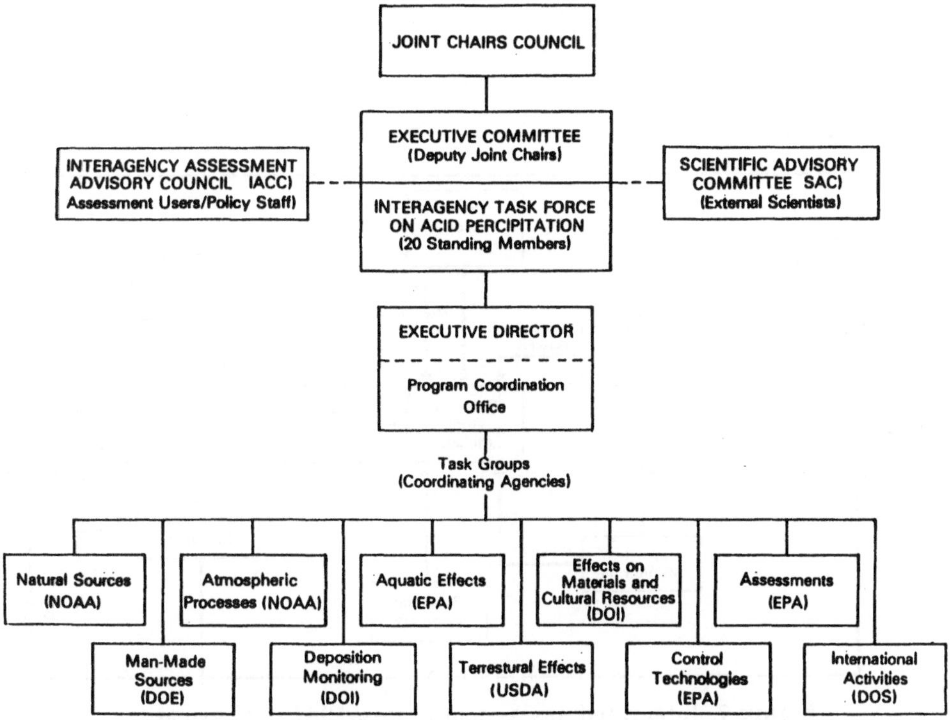

Figure 5. National Acid Precipitation Assessment Program

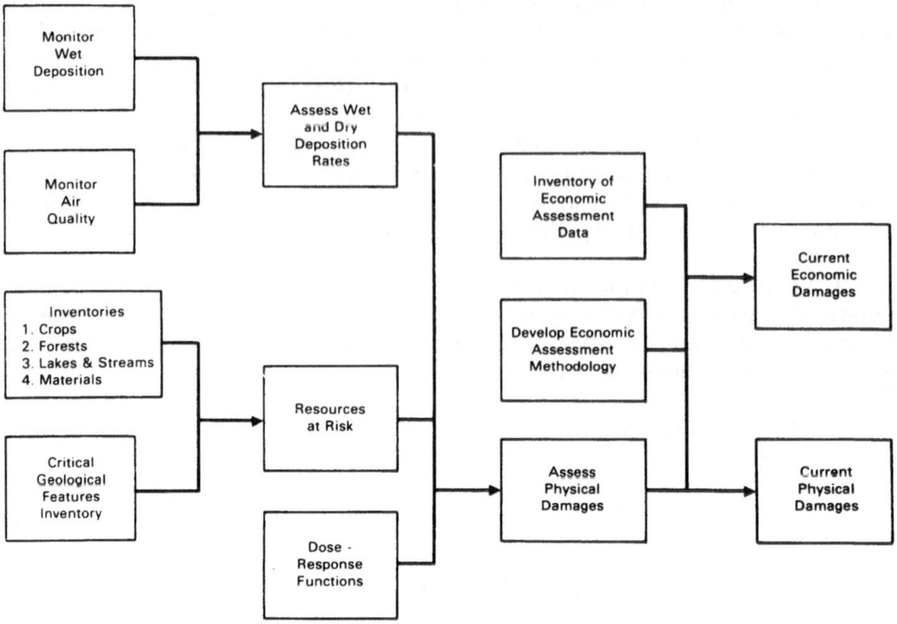

Figure 6. 1985 Damage Assessment by Task Group

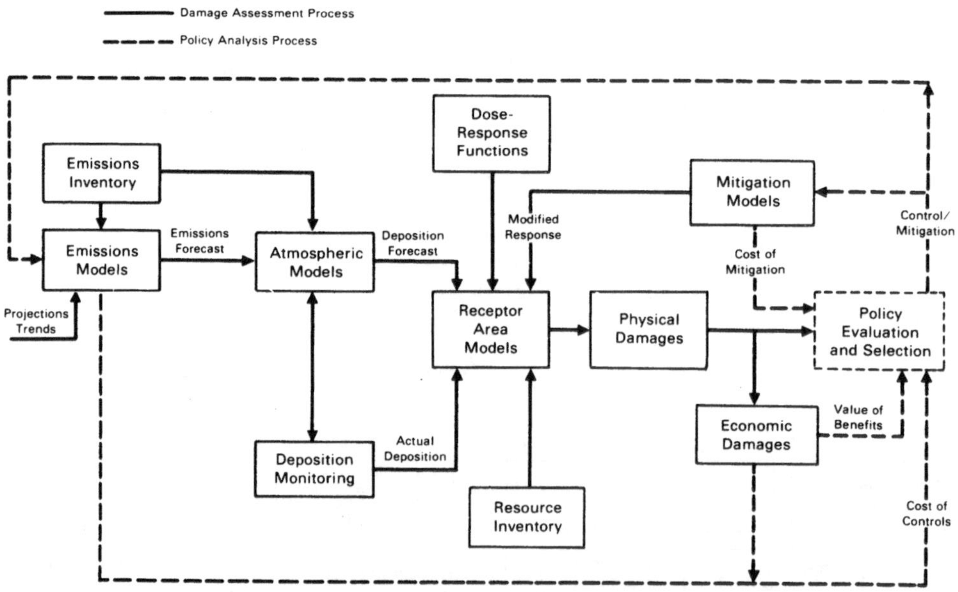

Figure 7. Integrated Assessment Process 1987 and Beyond

NAPAP is considering producing its first "integrated" assessment in 1987. This requires a framework that will incorporate the following elements into one comprehensive analysis: (a) changes in emissions and costs associated with alternative emissions strategies; (b) changes in deposition of substances in receptor regions of interest; (c) changes in effects related to changes in deposition; (d) costs and impacts of mitigation strategies; (e) economic value of changes in effects; (f) systematic uncertainty analysis; and (g) methods for conveying the results of this analysis to policymakers and interested parties. The 1987 report will incorporate all of these elements and relationships. The analysis may still be limited in coverage and completeness. Data and scientific relationships may not be available to include all relevant pollutants, types of effects, regions, etc., in all steps of the assessment. The 1987 Assessment will, however, at least demonstrate a methodology which links the best available information in all of the above areas in a consistent framework. This is distinctly different from the type of assessment to be produced in 1985. The 1985 Assessment will consist of results of analyses carried out in several of the relevant components of the program, which will be packaged together after their completion. Because of major gaps in available scientific information and a lack of a suitable framework, there will be no attempt to undertake an integrated assessment.

In addition to being modular rather than integrated, the 1985 Assessment will be regional rather than national. We do not, at the present, have the necessary information to perform a national level assessment. However, the 1987 Assessment will be national in scope. Also, the 1987 Assessment will focus on the benefits and costs of options to reduce or mitigate acid deposition. In contrast, the 1985 Assessment is to focus on damages to date from acid deposition. These damages are to be expressed in both economic and physical units. Damages to date can be interpreted several ways. A strict interpretation would require a careful, detailed appraisal of how the world would have evolved without acid deposition and comparing it with the current situation. NAPAP has opted for a more modest interpretation: estimating damages assuming that, with the exception of acid deposition damages, the world and society would have evolved the same way both with and without acid deposition.

The important differences between the 1985 and subsequent assessments are summarized below.

1985 Assessment		1987/1989 Assessments
Regional	vs.	National
Modular	vs.	Integrated
Damage to Date	vs.	Benefits Linked to Policies

RESPONSE TO THE NAPAP REPORT

Carl Griffith

Senior Economist
Policy and Planning Branch
Ontario Ministry of the Environment
Toronto, Ontario, M4V 1P5

Let me begin by stating that I am in complete agreement with the objectives of the NAPAP program and for the most part endorse the specific methodologies being applied to the assessment of abatement costs and to the estimation of acid rain control strategy benefits. Moreover, Mr. Currie has presented cogent arguments concerning the usefulness of economic analyses and their results to program development and policy making.

The sheer size of the program however, prevents a detailed response at this time. Consequently, my comments will address general issues.

TIMING

The first integrated assessment is due in 1987, subject to the qualifier that owing to insufficient data this assessment may demonstrate methodology, rather than present results. This in turn, implies that useable results would not be available until sometime after 1987. I wonder whether this time frame is consistent with the policymakers' requirements. If it is, does it imply that no decision to control acid rain will be forthcoming until at least 1987?

INFORMATION REQUIREMENTS

The 1985 assessment is to focus on damages from acid deposition and economic valuation of those damages. According to the authors, the output from this assessment must satisfy the decisionmakers' information requirements, namely, accurate defensible estimates of current and expected monetary damages. These requirements may be more ideal than realistic.

As you may recall from the previous discussion, what the scientist and the economist define as accurate or defensible estimates are seldom the same. My concern is that I doubt whether the scientific community would agree that accurate, defensible dose-response relationships are available upon which to base the 1985 assessment. And, if the economist's argument that defensible estimates do exist prevails, and the

1985 assessment proceeds in the absence of support from the scientific community, would policymakers be willing to act?

The authors did state that if appropriate information was not available a qualitative approach might be necessary. I would be interested to know more about this qualitative approach and whether its application would satisfy the decisionmakers' information requirements.

UNCERTAINTY

The recognition of the need to identify explicitly the uncertainties associated with various acid precipitation effects and with the relevant damages and benefits is well taken. There are, however, two uncertainty issues which should be highlighted: (1) the level of uncertainty of results; and (2) the communication of the uncertainty.

On the former, it was noted at several places in the paper that systematic uncertainty analyses will be undertaken, yet there was no elaboration of the nature of these analyses. Furthermore, since each analytical component of the program has some degree of uncertainty associated with it I am extremely interested in how all of the uncertainties are going to be characterized and assimilated into the integrated assessment. For example, the dose-response relationships may have an uncertainty of 20 percent, atmospheric modelling 30 percent, cost estimates 20 percent and economic valuation 20 percent. When all of the uncertainties have been added, will the variability of the results be so large as to render them useless for policy?

Assuming that the uncertainty issue can be suitably addressed, communicating results to individuals who are used to dealing with single point estimates will be a formidable task. Whether policy makers will accept or understand results couched in probabilistic terms is suspect. And, if ranges are used, my experience has been that the mid-value is adopted which often negates the purpose of presenting ranges in the first place.

COST MODEL COMPONENT

The authors indicate that the cost model will enable the identification of "least cost" strategies to achieve specific emission levels.

While I support the "least cost" efficiency criterion in strategy development, I disagree that the objective is to achieve a specific emission level. I would argue that the objective is to protect the environment by reducing acidic deposition in sensitive areas. As a result, we should be searching for control strategies which are least cost in terms of achieving some desired deposition level rather than an emission level.

Finally, I would like to make a general remark on modelling. The complexity and scope of your program could have two undesirable effects on policymakers.

First, if policymakers cannot comprehend how the results on which they must base their decisions were created, you run the very real risk of having them rejected, thereby omitting valuable information from the policymaking process.

Conversely, the modelling may act as an ameliorative drug upon which the decisionmakers become dependent, unquestioningly accepting what the model tells them, thereby alienating themselves from participating in the policymaking process.

In closing, the economic component of NAPAP seems to be an ambitious and considered effort to generate policy-relevant information. Having been involved in a similar exercise I empathize with the problems you face, but look forward to reviewing the results.

CONCLUDING REMARKS

SUMMING UP: AREAS OF AGREEMENT AND THE NEED FOR FURTHER RESEARCH

Allen V. Kneese

Senior Fellow
Resources for the Future
1755 Massachusetts Avenue, N.W.
Washington, DC 20036

I thought the opening discussion focusing on the natural science aspects of acid deposition was very useful, at least to me, since I did not know much about the science of it. I thought that one of the things that was of very great interest was the importance of dry deposition in this whole question. I had known that dry deposition existed but I did not know that it was, relatively speaking, so important. The discussion made me wonder about what is known concerning the relative damaging effects of wet versus dry deposition. Another thing that was interesting to me in that discussion was the role of oxidant-limited acid formation. This raised in my mind the question of whether we are really going to get what we think we are going to get if we make just a single-minded attack on sulfur.

I am going to make another few comments of a perspective nature and then I'll go on to the question of whether economic analysis is useful on the acid question. Freeman and some others have said that we should view the problem really as being one of long-range transport of pollutants rather than acid rain alone. The reason for this is that there are ambient effects associated with long-range transport of pollutants as well as with deposition. Frequent ones mentioned are health effects and effects on visibility. Then of course there are the effects of acid deposition, on materials, on ecosystems, on crops, all of those things which have been discussed, probably some of which are quite large.

THE RELATIVE IMPORTANCE OF NON-HEALTH EFFECTS

I wanted also to comment a little bit upon the health issue, because I am one of those who Lester Lave and Dick Wilson referred to as being inclined not to think that health effects are very large. If that view proves to be correct, then it has very large implications since, as was indicated in the earlier discussion, our basic laws are premised on the idea that there are relatively enormous health effects from air pollution. But I wonder about that for several reasons. One is that the macro-epidemiology studies that have been done over the last 15 years, simply have not come up with consistent results. Some have found effects and others have not found effects with respect to chronic morbidity and mortality; I am not referring to acute effects.

Furthermore, a recent study, whose authors had access to much better data than any of the earlier ones had, was a micro-level study, which found no chronic effects from sulfates, nitrogen oxides or particulates. This study was directed by my associate Paul Portney. This result cannot be considered conclusive because the epidemiological studies all have great deficiencies. Therefore, it is certainly arguable that chronic health effects exist, but I do not think that we have a research basis for asserting that they do.

Also, with reference to Dick Wilson's remarks, I don't find the comparisons of Japanese experience and smogs in London at all salient. The levels of exposure there were just so greatly more than the ambient conditions that we have in this country that I think it would be extremely dangerous to try to extrapolate the effect to our situation. I was in Japan as a consultant just before the Japanese decided to go full force into controlling their air pollution, and conditions were terrible there. One of the reasons for the terrible situation is that unlike the U.S., where we tend to have our industrial areas separate from our residential areas, in the industrial cities of Japan, industries are scattered all through the residential areas. I remember getting up in the morning in Fuji City, opening the shutters of my hotel room and seeing a petrochemical plant flaring off gases right outside the window. The situation was really much more extreme than it is here. I understand that since then they have done a great job of cleaning up.

Just one quick anecdote about the way they do things in Japan. I was there at the invitation of Professor Shigeto Tsuru, who is a very well-known Japanese economist and statesman. I was consulting about questions of policy with respect to pollution. On one occasion I was introduced to a group of businessmen who wanted to have lunch with me. I had thought those businessmen would be environmental officials of companies seeking information about environmental affairs in the U.S. However, the gentlemen were the presidents of Japan Airlines, Sony, Mitsubishi Heavy Industry, and Fuji Bank, etc. They were, in fact, the Board of Directors of Japan, sitting there in that room making a decision as to whether they should proceed now with massive pollution control. By then Japan had experienced unquestioned health effects.

What would be the implications, however, if it were to be true that these health effects may not be large in the United States? It could mean that we would have to re-think the basis for much of our air pollution law. It would then become very important to begin to understand, with some accuracy, what the other kinds of damages from air pollution are. We have suggestive evidence, for example, that visibility effects are important. I am personally persuaded that material damages are very large. Possibly broader ecological effects are important. Our basic law has been hung on supposed health effects, and if that supposition no longer holds up we'd better try to develop a solid basis for understanding what other type of damages imply for public policy.

RISK VALUATION AND RISK MANAGEMENT IN TANDEM

Another question of perspective relates to the matter of the separation of risk assessment on one hand, and risk valuation or management on the other. The need for this separation was mentioned by several speakers, and I agree that they are different activities. But it seems to me that they have to be done in tandem if they are going to be successful. At least they have to be done with good information exchange, because so often, if they are not, the risk assessment step produces a result that just isn't useful for the risk management step. For example, the risk assessment step may produce a single risk number when, in fact, what is needed for the economic evaluation and management decisions is some sort of function that will tell you how risks change when you change emissions. So I don't really quite agree that those two activities should be pursued independently.

THE ROLE OF ECONOMIC ANALYSIS

I want to come now to the question of the usefulness of economic analysis in the acid rain debate. My answer is that economic analysis is useful. I think we have had some very good illustrations of that in the presentations in this meeting. Two major quantitative economic analyses were presented.

1. Costs of Control

The first of them is based primarily on Hoff's Stauffer's work with adaptations by Paul Portney and Ron Jonash and others. These analyses address the question of cost-effectiveness of the low-sulfur coal vs. scrubbers alternative for reducing sulfur emissions. It is not that all differences among analysts have been sorted out by these efforts. But I am convinced that a basis has been laid for further study and analysis, and I think that these efforts have already gone a long way in structuring the information that is available into decisionmaking and policymaking contexts. Two issues struck me as needing further resolution, as the participants were going through their analyses. One of them is the evaluation of the effects on specified communities of shifts in coal markets. This is really the consideration that makes the huge difference between Ron Jonash's results and the other results, and raises some questions about how much that type of effect might already be built into the assumptions about the prices of coal. For example, many western states have heavy severance taxes that are covered by the price of coal and it may be that they already account for the costs of dealing with those community and state programs that are dedicated to community effects. A further question is how much of transportation cost is already included in the price of the coal. I think, therefore, that there needs to be some further examination of whether there aren't possibilities that double counting is taking place. I am not asserting that there is but to my mind it is still a question.

Another issue related to that study is the question of understanding how public utilities respond to regulatory incentives of one kind or another. Most of the reasoning about the economics of pollution policy by economists has been done on the assumption that we are dealing with a competitive economy in which resources flow smoothly from one use to another and in which profit maximization is the decision criterion. Yet, in the area of acid deposition, we are dealing almost entirely with publicly regulated sources, public utilities, and there are questions about how they will respond to different types of regulation. For example, one of the speakers said that while economic models assume that utilities will want to adopt the lowest cost strategy and will buy low-sulfur coal if that is the lowest cost way to control sulfur emission, in fact, some utilities claimed that they would rather install scrubbers even if the low-sulfur coal alternative were cheaper. One wonders why that would be, and a point to be considered is whether the utilities think they are not going to have to pay for those scrubbers because they expect to be subsidized. Perhaps there is another reason, but the question is whether utilities will behave in the way a real cost minimizing analysis would suggest. It should be noted that utility decisions would also depend upon how public utility commissions treat investment in scrubbers, i.e., whether the investment becomes part of the rate base or not. The same point applies to discussion of emissions fees. If the nation were to try either to use emissions fees as an incentive for control of pollution, or as part of a program to collect funds to help pay for a reduction of pollution, questions would arise as to how the utilities and the commissions that regulate them would respond to this new kind of cost. It seems to me that one of the important things economists should do in the acid deposition area is to try to understand much better than we now do, how regulated industries are likely to behave in these new kinds of contexts.

2. Assessing Benefits

The other quantitative study I would like to comment upon is Crocker's work on acid rain damage. That study is certainly a rudimentary risk assessment and evaluation effort. There is no question whatsoever about that. However, I disagree with my old friend Paul MacAvoy. I don't think it is meaningless. On the contrary, I think that what Tom did is reasonable under the circumstances and that, at the very minimum, it provides some hints as to where the major damage areas may lie. The study thus points to the areas that one may want to take a very hard look at in the future. One of those areas I believe to be materials damage. If I were to guess, by looking into my crystal ball, I would say that when we finally understand the situation, materials damage will turn out to be the biggest damage from acid deposition and other air pollution. Maybe this feeling is reinforced by the fact that I had to replace the awnings on my house last year. The awnings person condemned the awning material, but I wonder whether it's the cloth or if something else is happening. An odd thing about materials damage is that at first glance it appears to be a pretty straight-forward kind of thing to assess since it deals mostly with marketed goods. But the efforts at quantification that have been made in the past have not come to grips with it. I hope that the NAPAP effort described in this meeting can make some progress on this point, and I hope it will also be possible in this effort to give attention to the matter of how one would design a research program that will really give answers that have some dependability.

Let me turn now quickly to some areas that I think could have benefitted from more discussion in this meeting. Two of them are among the areas of controversy that were listed by Rick Freeman in his comments. They have also been mentioned repeatedly but very briefly by a number of other speakers. One of them is the question of the role of intrinsic benefits, that is non-user benefits. This is important in the case of acid deposition because some of the kinds of things one would want to evaluate probably do have intrinsic value. There may be people who, for whatever reason, national pride, a sense of propriety, or whatever, prefer to have environmental quality maintained, even if they do not ever themselves intend to use the area affected. Some of the contingent valuation studies that have already been done, have tried to get a handle on intrinsic benefits. The results are extremely uncertain and I wouldn't put much store in their accuracy, but I do believe that they have conclusively shown that such benefits exist in many instances, and that in some cases they may be quite large.

The second area that I think needed more discussion is the validity of contingent valuation methods themselves. Any large study of the acid rain deposition problem will probably have to involve contingent valuation methods and in many cases, also an assessment of intrinsic benefits. One may therefore use the intrinsic benefits question to illuminate what some of the problems are by doing a dependable contingent valuation. Let me cite an instance. A study was done on the value of protecting the Grand Canyon from haze which might be due to energy development in the Southwest region. One of the thrusts of the study was to get at the question of intrinsic benefits, so not only were visitors to the Canyon region interviewed and considered, interviews were also conducted in Albuquerque, Denver, and in Chicago, and included non-users as well as users. One of the really interesting results of those interviews was that the people in Chicago placed almost as high a value on protecting the Grand Canyon from haze as the people in Denver did, even though people in Chicago are much less likely to visit the Grand Canyon than are people in Denver. The initial reaction was that the intrinsic values must be enormous, and projections of such values were made to the national population and the result was enormous. Follow-up studies were then conducted. Similar questions were asked of small samples of people. However, before the question on evaluation of the Grand Canyon air quality was asked, respondents were asked to value some environmental benefit closer to home. Following that they were asked about the Grand Canyon. The result was that the "closer to home questions" in the Denver case didn't change the estimate for the Grand Canyon by much, but the estimate for the Grand Canyon dropped dramatically for Chicagoans. This type of result, as well as other evidence, has led people to believe that persons may have a mental account for various categories of market and non-market goods, including the environment. That is to say, they don't actually mentally process all the kinds of environmental issues that might come up, but nonetheless have at least some kind of a vague idea about how much they might be willing to expend on environmental goods. Furthermore, the argument goes, if you ask persons about one environmental item that is rather dramatic, they may give you the whole amount in their environmental account, but when it is brought explicitly to their attention that there are other environmental items which might also be valuable to them, then it comes home that a budget allocation has to be made. Accordingly, the newer contingent valuation studies are trying to design this kind of budgetary consideration into the interview instrument. But it's still an art and is one of the areas which, despite the great deal of research in recent times, still needs vigorous development if we are to be able to evaluate the kinds of air pollution effects that have been discussed here.

The third area that I think could have benefitted from additional discussion is the role of economic analysis in distributional considerations. A number of people have called attention to the importance of distribution aspects in the case of acid deposition and I don't have to repeat what they are, they exist for both the cost side and the benefits side of control proposals. A lot of the discussion here could have been read to mean that economics is interested only in efficiency and not in the distributional aspects. Indeed, economics is interested in efficiency and that may be its most important role in all of this. While politics is about distribution and appropriately so, I think it is extremely important, almost like a moral mandate, that economists call attention to the efficiency losses that are associated with certain types of public actions that are taken. But there are also things that economists can do to help make more informed decisions about the distributional questions. Indeed there have been some efforts to break out of the utilitarian framework of benefit-cost analysis and consider the economics of distributional questions from the point of view of other ethical constructs.

In conclusion, I want to say that both in reading the NAPAP paper and in listening to Currie's discussion of it today, it struck me that it's an extraordinary reasonable kind of document and approach. I hope this effort will enjoy great success and I hope that if we meet again on the economics of acid deposition, in three years, that we will, because of this enterprise, have a great deal more information than we do now.

DINNER SPEAKERS AND GREETINGS

NEW YORK'S STRATEGY FOR CONTROLLING ACID RAIN

Henry G. Williams*

Commissioner
New York State Department of Conservation
50 Wolf Road
Albany, NY 12233

Governor Cuomo is an articulate and eloquent spokesman, and is interested not only in the reduction of the acid rain menace, but in other causes related to the environment, to the area of social justice, and many other concerns as well. I am indeed intimidated to have it thought that I might be substituting for Governor Cuomo. Obviously, I cannot do that, and neither am I going to be delivering Governor Cuomo's address. What I will say is that Governor Cuomo does have an abiding interest in the subject which brings us together. Governor Cuomo talked with Woody Cole, Chairman of the Adirondack Park Agency, and the Governor repeated his concern that he could not be here with you today, but, when the New York State Legislature is in session, the Governor dare not be far away from Albany.

I am going to give my own kind of remarks. What I will do is focus upon some of the experiences that we in New York State have had with the subject of acid rain, in the hope that our experience and our interest may be of some value to you who come from the various places and representing the various kinds of interests that you do. We all know, of course, that acid rain is a severe challenge because it is a problem which no single state or single province can resolve by itself. We know that acid rain affects our environment in a variety of ways that we do not fully understand. New York State has major interests in acid rain. I'd like to share some of these experiences with you today.

In New York, we know that we have observed great damage in our Adirondack mountain region, for over 200 lakes and ponds are too contaminated now to support any fish, and hundreds more of our lakes are in danger of becoming that acidified. The damage hasn't been confined solely to our Adirondack region. Our Hudson Highlands and our Catskill Mountains are also suffering. Precious resources which have been preserved over many generations, oftentimes at great effort and at great expense, are being put in jeopardy.

*Commissioner Henry G. Williams represented New York's Governor Mario Cuomo at the conference.

Since 1968, New York State has reduced its sulfur dioxide emissions by nearly 50 percent, or nearly a million tons. We have applied and enforced some of the strongest air pollution control regulations in North America. Governor Cuomo, just earlier this year on August the 6th, signed into law a State Acid Deposition Control Act. This act mandates reductions in sulfur dioxide emissions by 1988, and provides further means of control thereafter. This law requires us to identify sensitive receptor areas. Having identified these sensitive receptor areas, we then establish environmental threshold values for them, and will prepare a program to achieve the appropriate reductions in the sulfates at these sensitive places. New York's Acid Deposition Control Law is, we believe, the first of its type in the nation. Several states have indicated to us their interest in the approach we have taken. Concurrent with the enactment of the Acid Deposition Control Law, we also developed a statewide sulfur dioxide emissions policy. In the environmental impact study prepared in association with this policy, a basic format is set forth which will enable us to fulfill many of the requirements under the new Acid Deposition Control Law.

All of this adds up to a demonstration of the substantial investment by New Yorkers. We have been studying, we have been acting, and we have been spending. Without accompanying national action, however, New York or any other state or province, cannot expect to reverse or offset the adverse impacts being inflicted by others. Our efforts are a start, but they are just a start. They demonstrate our intention to do our utmost to minimize our own contributions to the acid rain problem. We believe the federal government in the United States should accept its responsibilities for protecting New York and other states and provinces from the impacts of interstate pollution. Before I came in today, I was informed that the Environmental Protection Agency had just conducted a press conference to announce that agency's decision regarding the petition by New York State under Section 126 of the Clean Air Act. Under that petition we sought relief through the Clean Air Act to help protect us from interstate pollution. EPA is denying that petition, and continuing the present situation that we have.

Let me just say a couple of words about how the Acid Deposition Control Act will work. The regulatory approach to meet the requirements of the Act is currently being developed, and we expect that we will be having public hearings and providing the opportunity for general public discussion early next year. While we do not have all of the specifics at this particular time, we already know that the regulations will incorporate some basic principles, such as new rules regarding the allowable amount of sulfur in fuel. For example, a source that is currently allowed to use 2-percent sulfur fuel, may be required to cut to $1\frac{1}{2}$-percent sulfur fuel, depending upon the particular impacts of that source on the environment. The law will also provide for a system of offsets or tradeoffs that are based upon deposition effects, and these are an integral part of the Act. These offsets could allow a source to use a higher sulfur fuel, provided that other sources reduced emissions to provide total reduction in sulfate deposition. Under this new act we would also allow for new stationary sources to be built without requirements for these offsets if the new source meets federal new source performance standards, or uses best available control technology where new source performance standards have not yet been established.

All this is possible, as a result of the generic statewide sulfur emissions policy which we had previously developed. This is the draft of the environmental impact statement describing this approach; we believe it is a novel and creative approach. The comprehensive review of sulfur deposition all across the state enables us to take into account the concept of offsets, which represents the concept of balance, and to recognize that we are after a total load reduction. The Act also provides for the development of a nitrogen oxide standard and subsequent regulations related to the nitrogen sources.

The purpose of this Conference is the economic assessment of acid rain impacts. Certainly this aspect of the acid rain issue has been an absorbing matter of attention for a long period of time. We have heard various views of the cost of reducing sulfur emissions. Some say the cost is disproportionate to the benefits. Some say we lack sufficient understanding of cost-benefit relationships. Some say that costs will escalate if control programs are not started soon. But the costs of acid rain should not be measured solely as regulatory costs. We know that there are many other costs associated with the environmental damages that may reasonably be attributed to the effects of acid rain. Our studies in New York indicate that the damage which acid rain has done, exclusive of public health and man-made facilities damages, amounts to about 1.6 billion dollars. There are added indirect economic impacts that are associated with acid rain. For example, there is a perception suggesting the Adirondacks have become wasteland, a wasteland as a result of this thing called acid rain. And as a result, some people are saying, "Shall I go to the Adirondacks to enjoy myself--should I go to the Adirondacks to fish, shall I go to enjoy other aspects of that tremendous resource environment?"

Just before I left my office this morning, I bumped into one of our outstanding outdoor writers in New York State, a fellow who writes about hunting and fishing in the local newspapers. I told him where I was going, and he said to tell you about this point that I just made and that the Adirondacks are still a great place. While we have identified specific damage, and Woody Cole will tell you where that specific damage is, the Adirondacks are still a great place.

A challenge before us is to offset the perception that acid rain has caused a great devastation across the land. This is a cost; this is a cost that extends over a broad range, rather than a direct cost itself. What will be the cost, however, if forest decline continues? What will be the cost in terms of human health? Undoubtedly, the economic impacts recorded to date, combined with the potential economic losses, far outweigh the cost of regulation and control. New York State is bearing its share of the immense cost associated with acid rain. We are paying the highest utility rates in North America, and we have committed to even further burdens in order to do all that we reasonably can to reduce the damage which acid rain has and will cause to our resources, our buildings and the health of our people. The State of New York has done its part, and we now seek the equity of a national solution to the problem.

While various proposals have been advanced in Congress and elsewhere, no approach that I am aware of, other than New York State's Acid Deposition Control Act has been enacted to bring about specific, meaningful reductions in law in the emissions of sulfur dioxide. Governor Cuomo, in testimony before the House Subcommittee on Health and the Environment, advocated a nationwide 12-million ton reduction in sulfur dioxide emissions by the year 1995, and this is the same goal called for by the National Academy of Sciences, as you are aware. Governor Cuomo further recommended that the imposition of a tax on emissions of sulfur dioxide and nitrogen oxide be established as opposed to a kilowatt tax on non-nuclear electricity. Such a method of funding emission reductions would place the burden for reducing pollution on the polluter and that is where we believe it belongs. Governor Cuomo made it very clear that the 12-million ton reduction and the tax suggested were starting points for discussion and that he was prepared to negotiate to bring about the kind of reduction targets and funding methods that would reduce overall load. And in order to reconcile the different interests, the important thing is to achieve emission reductions, and to do so in a way that is fair among the states and provinces. New York wants to be a part of a solution, a fair solution, to the problem of acid rain. We believe we have done our share, and we have borne more than our share of the costs.

This conference will clearly be helpful in giving us a better understanding of the important economic aspects of the acid rain problem. When we finish our deliberations here, let us apply our added insights and knowledge to the central task to control acid rain, to spread the cost burdens fairly among the states and provinces. Thank you very much.

DISCUSSION

QUESTION: Commissioner Williams, assuming, as you expect, that the EPA does turn down New York State's petition, what is your reaction? Is there some problem between the federal government and the states right now? Is there some problem with the Reagan Administration?

COMMISSIONER WILLIAMS: Is there some problem between the Reagan Administration and the states, the gentlemen says! You might say that there is a little bit of a problem--yes. The Attorney General of New York State has been pursuing the §126 litigation, and I expect that we will be disappointed, perhaps not surprised, but disappointed, in the action that EPA is expected to take today. I am sure that what we will do is to re-examine the options before us, and to determine what the basis of the decision of EPA might have been. It is my understanding that what EPA will be saying is that they are denying our petition on the basis of insufficient evidence, or something of that sort. We will be making a determination as to whether or not such evidence can be compiled and will proceed, perhaps legislatively, I don't know.

QUESTION: How far back will that action set New York State in trying to clean up the problem, reviving those dead lakes in the Adirondacks?

COMMISSIONER WILLIAMS: Well, as I indicated, for a long period of time New York has not restricted itself to talking and asking, but we have spent our hard-earned dollars to focus upon ways in which we can address the problem. New York wanted to be in a position before the quorum of all the states in the United States, to be able to say that we are not just calling upon others to bear the burden. What we're doing is demonstrating that we are doing all we can to reduce the burden that we create ourselves. And that's why this Acid Deposition Control Act has been enacted. I might also point out that in addition, with respect to the generic study of sulfur dioxide policy, we believe this is a pioneering effort that represents a way to look at the total SO_2 problem on a statewide basis. It provides for us a clear and comprehensive view as to how we can deal with the question.

In addition, we have substantial other ventures underway. Woody Cole, the Chairman of the Adirondack Park Agency, has been leading the effort in his role as the chairman of that important agency. We have a 4½-million dollar project underway in cooperation with the Electric Power Research Institute, and there are many other ventures that have been accomplished. I would like to make two key points--one is that as a result of the enactment and the implementation of a stringent air pollution control legislation over a period of nearly two decades, New York has reduced the volume of sulfur dioxide that it contributes by one half, that is, by a million tons. But that has not come without a price. As I said in my remarks, New Yorkers pay the highest utility rates in the nation, and that is a very significant cost in terms of our economy. That is why I said in my remarks that there is another key aspect of the issue, and that is to share the burden in an equitable way.

QUESTION: Just to follow up my question--the question was, "How far back will a decision against you by the EPA set New York's efforts to clean up ?"

COMMISSIONER WILLIAMS: It won't set us back; what it will do is retard our forward progress.

QUESTION: By how much?

COMMISSIONER WILLIAMS: By how much? It all depends upon what the exact terms of the EPA decision happens to be. But as I have indicated, we have a number of ventures that are before us. These are financed by state funds, and they will go forward rather rapidly.

QUESTION: New York State has also been addressing the issue of nuclear power, and has been attempting to keep nuclear plants out. There is a perception by some people, that instead of benefitting the State of New York, perhaps you are not. Perhaps what you are doing actually is limiting economic viability for the lower income population of the state by raising utility costs, which will severely impact your economic well-being, while you are simultaneously limiting your tax base, because the total costs of the acid rain strategy will drive industry out of the state. Does New York State take no responsibility?

COMMISSIONER WILLIAMS: The question is, if we had been more supportive of nuclear energy, would we have produced utility rates that would be more attractive? I would call your attention to a major facility that is located on the north shore of Long Island at a place called Shoreham. I was with Governor Cuomo last Friday near that plant. This is a plant which started at a projected cost of less than half a billion dollars and ends up now with close to four billion dollars or more; it is a plant about which I heard person after person come forward and indicate that addition of that plant to our system would have tremendous impact on their business. There are two other key issues that Governor Cuomo has made very plain. One is the issue of safety, the questions of an evacuation plan to insure that our citizens will be adequately protected, particularly those citizens who live in a place like Long Island from which all egress points go through one narrow channel. The other question is, what will we do with nuclear waste? This is an issue that continues to be before us. And while the federal government has been studying this over a period of time, the fact is that a decision regarding the long-term disposal of high-level nuclear waste has not yet been adequately resolved. So these are fundamental questions that have established, as a point of policy in New York State, opposition to further nuclear development.

QUESTION: I recently received a copy from your agency of a study made of over a thousand lakes and streams in New York. My recollection is that most of the ponds and lakes that are acidified tend to be smaller than the average area and also at relatively high elevations, leading to the supposition that a lot of these indications are not representative.

COMMISSIONER WILLIAMS: My recollection of that study demonstrated the impact on the Hudson Highlands, which is east of Albany, the capital city, and the impact on the Catskills. The particular focus of attention has been on the western edge of the Adirondacks and has been fairly concentrated. This seems to be in accord with the models that we have that show at least the location for the spread of acid rain in relation to the sources. You really ought to learn more about the draft environmental impact statement, because it does contain the results of our extensive modeling work, to indicate both the source sites as well as what happens at the receptor sites. Thank you very much.

DECISIONMAKING AT EPA

Charles Elkins

Assistant Administrator for Air & Radiation
U.S. Environmental Protection Agency
401 M Street, S.W.
Washington, DC 20460

It's an honor to be here at the symposium sponsored by your organization, which has a national reputation for providing information on a variety of subjects in a setting in which all parties can participate openly and with a fair hearing. I am glad to be here tonight. I apologize for the fact that Bill Ruckelshaus, Administrator of EPA, is not able to join us tonight. I know I don't have to tell you that he has a very strong professional commitment to finding solutions to the acid rain problem, and he has devoted a great deal of his personal time to that effort.

I may be able to give you some insight about what goes on inside the executive branch where decisionmaking in areas like acid rain goes on and to try to give you a feel for the kind of role that quantitative analysis plays in making decisions in a field like acid rain. Now, all of you experts out there, please keep me on the straight and narrow!

One of my jobs in this acid rain business is to try to keep track of what is being said about acid rain around the country. I read newspapers for this, and am amazed every day about how simple the acid rain problem really is. If you read the Boston Globe you know that delay is unreasonable and that we really should proceed expeditiously to deal with acid rain. And if you read the newspaper from Charleston, West Virginia, it is equally clear that we should not proceed at this time in light of all we don't know about the acid rain problem; in fact we do not even know if there is an acid rain problem. So, I am gratified to talk to a conference such as this, where we're focused on quantitative analysis and are not thinking that the issue is so simple that we can readily tell people the answer. If we knew what the answer was, we wouldn't need all this quantitative analysis.

I would like to suggest for your consideration that the people who do have a quick answer on acid rain, and I'm talking here about the people who just participate in a casual way, not those who really are into it in a more serious manner, that they may have too simple a decision model in mind as they address this problem. I think their decision model is something like this. It is an "if-then" formula. "If there is a serious environmental problem with regard to acid rain, then the government should really take steps to solve it now." If you use that formula, then acid rain is a very simple issue. But if you are not going to use that simple approach to making a decision about acid rain, what model should you use? There are probably a thousand choices, but let me suggest a decision model or how you would approach making decisions about acid rain.

This model would have three parts. One part identifies the nature of the problem and the degree of certainty which we have with respect to those statements about that problem. A second part asks what the possible solutions to the problem are and discusses how certain we are that the solutions will actually solve the problem we just identified. Finally, the third part is concerned with the impact of these solutions on various parts of society, and the degree of uncertainty we have with respect to those impacts. Now you notice that, for each part of this decision model, I mention uncertainty, which is a very important element of this decision process. We should confront it and deal with it. You can't wish it away; it is there; it always will be there. By the same token, we shouldn't be intimidated by it. We face uncertainty every day at EPA in many ways; in fact it is sort of a token of our line of work. EPA, in a sense, is like an insurance company. We, as professional staff members, assess various risks. We ask ourselves what we know about these risks, how we can deal with them, and in many cases, in essence, what we do is sell insurance to the American public. When you issue insurance policies, whether you are an insurance company or the EPA, you don't need definitive answers. EPA is quite ready to deal with uncertainty. It's part of our everyday life and the real question is how to factor it into your decisionmaking.

Well, let's go through this model that I just suggested and try to apply it to the acid rain problem. We start with the nature of the problem and the uncertainties about that problem. Look first at the aquatics area. What do we know about the effects of acid rain on aquatic resources? We are fairly certain that there are lakes in the Adirondacks and probably in Canada which have been acidified by deposition. And you probably know that EPA is in the midst of conducting a survey of a representative set of lakes across the country to try to quantify that problem more exactly. We don't know at what rate those lakes became acidified but we do know from laboratory studies that sport fish have great difficulty surviving in lakes which are acidified below a pH of 5. We are fairly certain that sulfate plays a larger role in this process than nitrates. However, the spring runoff which is high in nitrates is certainly a factor that has not been fully assessed. We know we have a problem in the aquatics area; we think it is caused by deposition; we don't know the rate of it, but as problems go, that's not too bad a record of information.

We turn then to the other major area of concern at this time--to our forests. Here, we are much more uncertain. We don't know the extent of the problem, that is, whether it is only a high elevation problem in this country or if it extends beyond that, and perhaps even more importantly, we don't know the cause of this problem. Is it the acidity? Are we over-fertilizing these trees? Are we poisoning them with heavy metals? What about ozone? Is there some synergistic effect between these various pollutants that will make it very difficult for us to single out what's happening? Certainly we can agree, however, that if the forest issue continues the way it has for the last year, and we find that in fact we have a serious problem in this country, we can expect that it will play a much larger role in the decisionmaking in Congress and the executive branch than it has so far. Aquatics damage has been the main focus in the past, but ongoing research on forest effects could change the debate dramatically.

In summary, in the aquatics area, we have good, qualitative evidence of cause and effect but very little quantitative information. In the forest area, we have neither qualitative nor quantitative evidence.

In the second phase of this decision model, we look at possible solutions and the uncertainty that surrounded the effectiveness of the solutions. But, what do we find here? Starting first with the forest area, even if we determine that we do have a serious problem in this country, we don't know what the cause is, and even if we find the cause, will we know how to remedy it? Let us assume for instance that it is a heavy metal problem. Dr. Bruck who is doing research on Mt. Mitchell indicated in a speech the other day that he was finding very high levels of lead, copper and zinc in the soil at the top of Mt. Mitchell--250 parts per million of lead, as I remember. He then went on to say that, by his estimate, it would take 5,000 to 10,000 years for that soil to purge itself of that heavy metal, even if all emissions of lead were eliminated! Well, that would wipe out the idea that we might be able to do something right away if we were to find that lead deposited on the forest floor is the cause of our problem. I don't want to make light of this but it is a way of trying to illustrate the point that you might very well be able to determine what the problem is and what the cause is and still be stymied in figuring out what to do about it. It is not quite so difficult if the problem is nitrates but the trend in nitrogen emissions is upward, and it is not going to be easy to turn it around. So, these are uncertainties about designing a control program that would actually impact the problem as we identify it.

Let us turn to the aquatics area. Here, the conventional wisdom has not in our view kept pace with the science during the last year. And as I go around making speeches on acid rain, I don't seem to find anybody except myself who deals with this aspect of the subject, which is the rate at which we see the acidification process going on in watersheds. As I indicated, there is a strong concensus that in the past deposition has been the cause of acidification in a number of lakes. However, a control program would most likely be focused on preventing additional lakes from becoming acidic rather than merely trying to restore damaged lakes. How many lakes will be damaged in the coming years by deposition? The reason this is tricky is because, as you are quite aware, sulfur oxide emissions in this country have been coming down in a fairly dramatic way over the last ten years, 28% since 1973, in spite of the fact that there was major growth in electrical energy production during those years. That is not to say that 28% is good enough, but when you look at the projections for the next couple of decades, the emissions of sulfur oxide are expected to remain relatively level and, by the year 2000, we expect them to resume their trend downward although this is not certain. That means we have to ask the question, "What kind of damage will take place during the next few decades in the sensitive watersheds with deposition at the levels that we have today?" I think people feel that there may be a fairly simple answer to that. Simply measure the neutralizing capacity in the various watersheds and estimate in how many this neutralizing capability would be overridden by deposition during the next couple of decades.

Well, quite frankly, we used to think it was that simple too but during the last year-and-a-half, or so, with the output from the research program which we've received, we have begun to realize that we don't know as much as we thought we did. This is because the absorptive capacity of the soil in the watershed also plays a role in preventing acidification, and it is not easy to predict the absorptive capacity of watersheds and their vulnerability to acid deposition. Watersheds could be put into three classes. One kind of watershed has plenty of neutralizing and absorptive capacity so that we could continue to impact it with deposition for our lifetime and the lifetime of our children and we would not really expect to see much effect. This type of watershed would essentially be able to protect itself and, I think we could all agree, it should not be the target of a control program during the next couple of decades. A second type of watershed has very little absorptive capacity and a low neutralizing capacity and cannot protect itself against today's level of deposition. Now, you may think it is these watersheds we should be designing a control program to protect. The trouble with that reasoning, however, is that it is quite likely that, if they are in that condition, they have been in that condition for a while. Most of them have probably been acidified by the heavy levels of deposition which we have had over the last several decades in this country. We could hope to bring them back to a more natural state, and that should be emphasized. In terms of protecting them in the future, we are a little too late for that.

Then, there is a third category where there is a limited amount of absorptive and neutralizing capacity. If we keep putting the same level of deposition into these watershed lakes over the next couple of decades, their lakes and streams may become acidic. Now here's the $64,000 question. You're designing the control program; you need to tell your boss how many of those third-category lakes there are because that's the group you are going to tell the American people that you want to protect. How many third category lakes are there? We asked the National Academy of Science that question and they didn't know the answer. I'm not surprised. It is a tough question. They did give us some suggestions about how, over the next several years, we can make an attempt to find out how many watersheds are in that third category.

You might ask, "Can we afford to wait several years to find that out?" That depends on how fast we think this process is taking place in these watersheds. How fast do these watersheds come to the point that they can no longer protect themselves against what we are doing to the environment. Well, we don't know how fast that process is going on but, based on what little we do know, we have reason to believe that there are not too many lakes in this third category and that they are not changing very rapidly. A high priority of our research program is to find out as quickly as possible whether we are facing an emergency situation with regard to that third category of watersheds and whether they are a significant portion of our aquatic resources. In short, we wish to know the cost of delay. If it is great, we should proceed with the design and implementation of a control program.

Returning to the three-part decision model, we have looked at what the problem was, and what the uncertainties were. We then looked at potential solutions. I would present to you the proposition that we have considerable uncertainty that proceeding apace with a control program would not, in fact, produce the results that we would be promising the American public. Just because we know that we have a problem does not tell us how to design an effective control program.

Turning, finally, to the third category of the decision model--impacts of the solution--you have heard a lot about this subject today. I will not extend it, except simply to say that the cost of a program is certainly a relevant factor, and there is a lot of interaction between those impacts and the ideal design of the program. If you know that it is going to have a major impact on high-sulfur coal miners in the Midwest, then you may have the response that Congressman Waxman had, which was to seek a solution which would minimize those impacts. How certain are we about the potential impacts of the various proposed solutions? You could say you are fairly certain about these costs and these impacts, and I would have to admit that when you compare what we know about the impacts with what we know about the problem and the possible solutions, certainly we are much more sure of ourselves in the area of impacts. But, I must recite to you a case involving my boss a few years ago. He made a decision that all cattle with more than a certain concentration of a pesticide in their flesh be sacrificed. We estimated the total cost of killing all those cattle, and we put that down as part of the cost of the proposed solution. My boss subsequently left the government and a few years later was asked to come back and look at that decision just to see what really did happen. And lo and behold, he found that not a single cow was put to death. The people who owned those cows had found a way to fatten them up so that they spread the pesticides through the body and met the EPA tolerance level. It didn't cost a thing, except the money to fatten them up a little bit. Well, that was back in the old days when we didn't know how to do quantitative analysis, but never underestimate the ingenious efforts of businessmen who are faced with a problem. They will find the cheapest possible way to solve it, and they are often more creative than we analysts.

I have laid out a model that I suggest is a good one that you might want to adopt. However, it doesn't give the automatic answers of the "if-then" model that many people like to use today. Let me show you how it might be exercised in another case which may be more clear cut. Take the case of chlorofluorocarbons (CFCs). You may remember reading about this problem years back. We knew that fatal cancers are caused by people's exposure to sunlight, and we knew that the ozone layer serves as a protective barrier to that exposure. We knew that a decrease in the ozone layer would probably result in more cancers for the public and felt that CFCs were having an adverse effect on the ozone layer. When we looked at the solution to that problem we found that eliminating CFC use in aerosol cans would clearly reduce the emissions to the atmosphere but we weren't as certain about whether that change in CFC emissions would in fact have a significantly positive effect on the ozone layer. Looking at the costs of the solution, we found that the impact was quite narrowly focused. The people who produced those propellants certainly would be severely impacted but the larger industry could find substitutes for spray cans. And so, with a high degree of confidence about the fact that we could reduce the CFC emissions, could sustain the cost impact, and that ozone depletion would have fatal effects on Americans and people all over the world, but with uncertainty about the efficacy of our solution, we still took out an "insurance policy". We bought that insurance policy with the uncertainties we faced. The question before us on acid rain is, "Should we buy an insurance policy for acid rain as well?" If the premium is cheap, there would be no one in this room who would not suggest we buy the insurance policy. Obviously, the more expensive it is, the more you are obligated to think about whether you know enough to make that decision today.

That's where quantitative analysis comes in. This analysis can take the form of cost-benefit analysis, cost-effectiveness analysis, or whatever. The more we can quantify factors leading to decisions, the better decisions we can make. But I would like to suggest to you that in the acid rain area, the weakest link in this chain is not our ability to do cost-benefit analysis in the abstract sense, but rather the empirical data which must feed such an analysis. We in EPA are funding much of the work that is going on in quantitative analysis for acid rain, and we will continue to fund it because we think it is important for decisionmaking. But it would be a mistake to think that quantitative analysis would be the driving force or the sole determinant of the ultimate decision. In EPA we are developing a very complicated model that takes us all the way from emissions to the transport of these pollutants to deposition and all the way to effects. We are trying to quantify them in both dollar terms and any other way we can.

A question that I could pose to this group, based on what you have heard already in this conference, is, "How would you quantify in that model the value of the lakes which will be saved by the emissions reduction that you are costing out?" But that is not the issue; the issue is, how many lakes will be saved by the control program. This is obviously where it becomes vitally important to know how many lakes and streams will be acidified during the next few decades at current deposition levels. The same thing, of course, can be said in the case of the forests. We can place a value on the trees in U.S. forests--but how big is the threat and how much damage will be prevented? We can "solve" this by showing a range but how do we know that we have bounded the problem? We have to know enough about the science to be able to bound the problem. All of the proposed estimates of benefits from acid rain control suffer from this fatal flaw and therefore cannot be used to make decisions today.

Cost-benefit analysis, of course, is optimizing the greatest good for the greatest number of people. As you know, our country doesn't run on that principle; I'm not sure we would like it if it did run that way. There is a lot more involved in this issue of acid rain than trying to find the greatest good for the greatest number in this country. Acid rain is a classic case of an equity problem. The people in New England are angry that the people in the Midwest are ruining their lakes and threatening their other resources. The question is what is fair, not what is the cheapest solution. So, just because we in EPA will continue to fund quantitative analysis on acid rain doesn't mean that, as we go through this decision process, we will choose the option that comes out best in the cost-benefit analysis.

If you look to see how cost-benefit analysis and other quantitative analysis are used in decisionmaking at EPA and other agencies, I would suggest that, the higher you go within the organization and in the government, the less these analyses are used. It is an excellent way to identify those options that one wants to present to decisionmakers but, as you move higher in the organization, higher into the final decisionmaking, there are a lot of other factors that come to bear.

You may ask then how did EPA, or how did the President, or how did the Administration come to a conclusion in the acid rain area. Faced with all these uncertainties, Bill Ruckelshaus began to ask questions. Do we need to make the decision now? Is there a forcing event that really makes us choose now, based on what we know at this time? And what are the risks or the opportunity costs of delaying this decision until we can get more resolution on these issues? Now, if the situation is changing rapidly, if we are "going to hell in a hand basket", we need to know that so we can act. But, as we look at the science in the acid rain area, we don't find it likely that there is rapid change, and we don't find it likely that significant damage will take place during these next few years while we are accelerating the research program. We don't know this for sure, but we have proceeded to double the research program, with the highest priority on those pieces of research which will reveal if, in fact, we are faced with an emergency situation. At the same time, we are moving forward this year with 3 million dollars in grants to the states that Congress provided to us to look at the implementation issues related to an acid rain control program. We will be announcing these awards to the states within the next few weeks. We are not going to wait until we decide whether there should be a control program before we look at the difficult implementation issues of how the decision-makers in Ohio, for example, will divide up the emission reductions that they may well be faced with. Do they allocate them all to utilities? Do they look instead to assign some of it to industrial sources? Will they try to protect local coal? What are the facts that they need to make those decisions? This analytic process is going on this year with a number of studies.

This has brought us to the point of deciding not to proceed with a decision on acid rain at this time, but rather to do further research. This is not a comfortable position to be in. Quite frankly, this decision made everybody mad. The debate really focused around the proverbial, "Is the glass half full or is it half empty?". There are a lot of honorable and wise people who think the glass is at least half full, that there is no need to take any action, that the problem will solve itself, if there is a problem. The Clean Air Act will take care of it.

There is another group, equally honorable and wise, who believe the glass is half empty and we need to act now; to delay is to condemn a significant portion of our environment to destruction, and EPA is being irresponsible. So both groups think that we are wrong, but quite frankly we have chosen to call it the way we see it. We're right in the middle.

It's easier to take a guess and respond to the intense pressure than it is to take this position we have. We are not seeking definitive answers to these questions. We don't need 20-20 vision to make this decision. We are in the insurance business of taking risks and we are going to continue to be in that business.' But we feel in this case it is premature to buy the insurance. This position is politically the most vulnerable but we believe, ultimately the most sound.

DISCUSSION

JACKIE TUXILL (New Hampshire Citizen's Task Force on Acid Rain): I was wondering if you could talk about what you know with certainty that you are looking for, and when you might foresee reaching a level of certainty such that you can act. I would like to reference two reports. One is from the Office of Science and Technology Panel in June of 1983, which recommends immediate, meaningful reductions and action even without certainty. The Panel also expressed concern over irreversible effects from a delay. The other report is the OTA's report this summer which said that OTA did not perceive any real increase in scientific certainty in the near future and felt this was a policy problem, not a problem of science.

CHARLES ELKINS: That is the toughest question to answer and the reason is because we don't know how fast the science is actually going to provide the answers that we need. We are not asking for total information. We are looking for a little bit more guidance on how much of the resources are at risk, what the causes are and how we can deal with the problem. The question of how much certainty one should have in making that judgment is really a policy question; it is not a scientific question. The policy question is, how certain should the leaders of this country be about the statement of the problem and the effectiveness of solutions before they commit X-million dollars to that effect. This is a policy judgment, one that was made by the President of the U.S. "The buck stops here," as Harry Truman said, and it stopped on the President's desk in this case. You're really asking two questions, how soon will our research program produce results, and how soon will the President of the U.S. and the Congress of the U.S. decide that what they have is enough. It's a mistake to think that you have to wait several years before you can ask that question, and you ought to ask that question every single day. You look at the research and you ask, "Is this enough?" And that's what our job in EPA is and we will continue to ask that question and try to bring decisionmakers to the point of saying, "No, it's not enough," or "Yes, it is, and let's go."

DAVID LANTZ (Hoosiers for Economic Development): What is your general impression of the opinion that, when all is said and done, perhaps the wisest solution would be the one recommended by Mr. Ruckelshaus to the President a year ago, with respect to controlling the near sources as opposed to the distant sources?

CHARLES ELKINS: Mr. Ruckelshaus has not said what he recommended to the President. But he feels like all of us who work for bosses, that we ought to give our very best advice to our boss and let him make the decision. The option which the President chose is one of the options which he provided to the President. There are a number of people who feel that a solution which would be a relatively small decrease in emissions, such as something considerably short of 10 or 12 million tons, might well be a good interim solution. Let me suggest a couple of difficulties with that which, I think both environmental groups and businessmen would have trouble with. The first is that once you have gone forward with a 6-million ton reduction program, for example, if you want to wait until you see what that did for the environment, you've got a long wait. It would be eight to ten years to install the equipment, depending on how you try to control the emissions, and then probably five or more years to monitor the effects. It is hard to see that as a mid-course correction time in a two-phase program. More importantly, and I think this is the point that both environmentalists and industry might make, if you ask a company to make a six-million ton reduction and then a few years later you ask for another six-million ton reduction, a problem will arise if the response to the first reduction, which might be fuel switching, is inconsistent with the necessary response in the second phase, which might be scrubbing. So you end up with tremendous resistance against taking the second step. You are almost brought to the point, when considering a two-stage program which starts out with something smaller than what might ultimately be needed, of asking the businessman to take the steps which would be necessary for the full two stages, to avoid coming back a second time. Now if you will help me write that into some kind of law ... that would be difficult but I was trying to illustrate the point that it's not as simple as it sounds simply to take a small initial step and think that we can decide later to do more.

WOODY COLE (Adirondack Park Agency): Aren't we really talking about a problem of will rather than simply a problem of knowledge? I hark back to the Pure Waters

Program and the Clean Air Act. It was simply unknown by governments generally whether or not the Pure Waters program would end the eutrophication of lakes, or whether getting rid of the phosphates would have a good result, or if some of our major rivers and tributaries, such as the Ohio River, could ever possibly be clean again. I think you could talk about this in conjunction with air quality. Is it possible to make the air significantly cleaner by cutting off the sources of fly ash and hydrocarbon wastes which were dealt with in the Clean Air Act, and wasn't there a commitment at that point, a will on the part of this nation in the late '60's, to say that it is a basic premise that, as civilized Western humans in the United States, we clean up this planet earth, starting with ourselves? And now it seems the kind of thing that you are talking about, first try 6 million tons, then go back later for another six million tons, is simply a result of failure of will on the part, not only of the Administration but perhaps even of others in the country, to pursue a kind of program which starts with the premise that clean water and clean air are an absolute necessity. Is it not the fact that we have lost this impetus?

CHARLES ELKINS: I don't have a good answer to that because it's a political judgment that each of us has to individually make, but let me make some comments on it. I think that it is not that simple. I happened to have been involved in that eutrophication decision back in 1971. We almost made a major decision to take phosphates out of detergents. That would have been the equivalent, perhaps, of rushing forward right now on an acid rain solution. It wasn't a lack of commitment on whether we should clean up the water and deal with eutrophication; it was a matter of what was the best technique. And we decided, instead of taking the phosphates out of detergents, we would instead do watershed studies and then control. We found, of course, that sewage treatment plants are not the only source of eutrophication problems in these lakes, and we would have gone after half a solution, if we had controlled detergents. I'm not saying that it would not have contributed, but it would not have been the panacea that an awful lot of the American public were led to believe, unless we had done our science first. The commitment to clean air is there as evidenced by the fact that we, as American citizens, have spent 150 billion dollars over the last ten years to do something about clean air. The question is, do we know what to do in this case? This administration has committed itself that, when these uncertainties are further reduced, it will craft and support an acid rain solution. We can debate about what was really going on in Ronald Reagan's mind, and you can ask him if you get a chance, but short of that, you are asking, "How does the EPA deal with this problem?" I can address that. There is a very strong commitment on the part of the EPA and Administrator Ruckelshaus to solve this problem. But the more he has worked on this, the more he has sincerely concluded that it would be premature to launch a large control program, and I respect that judgment because he personally knows more about acid rain than anybody else in the Agency. He has personally come to grips with that problem and that's where he stands at this point. He could be wrong, and as I said, with the very high stakes of the game, I don't think it is really a matter of will.

CHARLES ELKINS (response to final question): Well I certainly agree with you that I have not covered all the important potential effects of acid rain. Certainly the research program and the decision process looks way beyond forest and aquatic areas. We are spending the money on researching some of these other effects because we feel, as you do, that some of those may in fact be the biggest cost items. I think I would disagree with you, however, about whether these other effects would drive the decision in the Congress, because I really think that that isn't the way the political process works. You can tell them, until you are blue in the face that, for example, the materials damage from sulfates is 3 times any cost you could find for aquatic damage, and I think they will yawn in your face. If you tell them about the lakes and the forests they would react--and I think that is not a silly reaction on their part. Materials are important in this country but not nearly as important in many ways, maybe not in the cost way but in many ways, to the American public as their lakes and their forests. And that is what I think is driving the issue and that is why I concentrated on those. On your second point, I am not sure I understood it, you're saying that, if industry is allowed to find the cheapest way, there are lots of different ways they will respond. Is that your point? I certainly agree and our administration has always advocated that that freedom be allowed because otherwise we are going to pay a tremendous cost to protect coal miners and surely there must be a better way to address this problem--not to downplay the importance of the impact on the coal miners. We're talking here about an incremental type of program. There are some real dangers to deciding on what looks like a relatively cheap, let's buy the cheapest tons first, kind of solution because if you really are serious in thinking that you may have to go to the second stage, you'd better think about how you are going to get there, because society may not be very eager to go with us to the second stage. Thank you very much.

ACID RAIN AND ECONOMIC ASSESSMENT

Richard L. Trumka

President
United Mine Workers of America
900 15th Street, N.W.
Washington, DC 20005

On behalf of the officers and members of the United Mine Workers of America, I would like to thank the Acid Rain Information Clearinghouse for inviting me to address you today. Before I focus on this afternoon's topic, let me give you a little background about my Union.

The UMWA is composed of approximately one-quarter million working, retired and unemployed coal miners throughout the United States and Canada. Relative to some other unions the UMWA is small in numbers, but we are the nation's largest energy producing union. Our members mine every type of coal that is produced in the United States, anthracite, bituminous, subbituminous and lignite. While our membership is somewhat concentrated in coal mining communities of Appalachia and the interior basin, we also represent thousands of surface and underground miners in the South, Southwest, northern Great Plains and the Rockies. All told, we represent about 70 percent of the workers in the U.S. coal mining industry.

Because our membership is spread throughout all of the nation's coal producing regions, virtually any issue that affects the coal industry has an impact on us. We do not share the luxury, however, of some segments of the coal industry. We cannot view issues in terms of East versus West or high-sulfur coal versus low-sulfur coal. To us, coal is coal whatever its composition or location. The paramount concern of the UMWA is the men and women who earn their living by mining coal. Thus, when we examine a public policy issue, we always do so in light of the potential effects on the people we represent.

That brings us to the subject of today's discussion: acid rain. Acid rain is an issue that has received a great deal of attention in recent years. The subject has been debated in the news media, in academic and scientific circles, in corporate boardrooms, in public forums, in state legislatures, in executive branch agencies, in Congress and in the courts. The intensity of this debate, and the voluminous amount of information that has resulted from it, point out the complex nature of the problem. All of us who try in vain to keep track of the evolving scientific knowledge of acid rain understand all too well the extreme complexity of that aspect of the debate. The politics of acid rain also has proved complicated. Due in large part to disputes over the efficacy of acid rain legislation, Congress has been unable to reauthorize the Clean Air Act since 1981. And it appears that the political debate will continue to be just as thorny in 1985 and beyond.

This week, however, our focus is on another complex aspect of the acid rain debate, the economic issues that arise out of control programs. According to the Council on Environmental Quality, our nation already spends in excess of $25 billion per year to control air pollution. By far, the bulk of this effort is aimed at control of pollution from electric power plants. Despite the high cost, most Americans will agree that the effort has been good for our nation both in terms of public health and the environment. We must recognize, however, that some of the costs associated with our cleanup efforts have been unnecessary and extreme. I am speaking particularly of the significant disruption of coal mining jobs and communities that occurred in the 1970s as utilities attempted to meet the original New Source Performance Standards. Because utility companies were "free to choose" any method of achieving the 1971 SO_2 standards, coal miners had to shoulder an excessive burden in our clean air effort; many of them lost their jobs. When the Clean Air Act was reauthorized in 1977, Congress agreed that existing policy placed too much of the cost on one segment of society and took steps to ensure that coal miners and their families would not be forced to sacrifice their livelihood as the nation pursues the worthy goal of a cleaner and more healthful environment.

Now we face a potential repetition of the economic dislocation that occurred a decade ago, but the magnitude of the dislocation promises to be far greater. Depending on the type of acid rain control legislation that might be adopted by Congress, as much as 150 million tons of annual coal production could be displaced. This would mean that 40,000 to 60,000 coal miners could find themselves without jobs. By any measure, that is a staggering number of displaced people. Let's look at how this might occur.

Many different acid rain control bills have been introduced in Congress, but two fundamental approaches have evolved. One approach simply mandates the level of sulfur dioxide reductions that must be achieved in a given area and sets deadlines for compliance. The means to achieve compliance is left entirely to the utilities. This approach is embodied in S.768, introduced by Senator Stafford of Vermont and adopted by the Senate Environment and Public Works Committee in March 1983. The other approach seeks to address the problem of significant electric rate increases in midwestern areas and employment disruptions in coal production states. This approach is embodied in H.R. 3400, introduced by Representatives Waxman and Sikorski and co-sponsored by over 100 members of the House.

Let's first examine the impact of a Senate-type bill on employment among coal miners. S.768 would require a 10-million ton reduction of sulfur dioxide emissions in 31 states adjacent to and east of the Mississippi River. Under this proposal, each utility would be free to choose its own methods of achieving its share of the mandated SO_2 rollback. Among the options that a utility might choose are: retiring older units and replacing them with less polluting ones; retrofitting plants with scrubbers; or switching from high-sulfur to low-sulfur coal. The degree of fuel switching that occurs as utilities attempt to reduce emissions will determine the extent of employment disruption among coal miners. Of course, we expect that under such legislation each utility would seek to minimize its direct cost under the program. Accordingly, we would expect a mixed bag of compliance strategies, retirement and retrofitting of some units and switching of fuel supplies at others.

The U.S. Department of Energy has estimated that between 50 and 75 percent of the emissions reductions required under a Senate-type bill would be achieved by fuel switching. If 50 percent of the reduction is met by fuel switching, we estimate that 40,000 coal miners in high-sulfur coal producing regions could lose their jobs. Most of these jobs would occur in the Appalachian coal producing states of Pennsylvania, Ohio and West Virginia and in the midwestern coal producing states of Illinois, Indiana and Kentucky.

If 75 percent of the emissions reductions are achieved by fuel switching, we project that nearly 60,000 coal miners could face the unemployment line. Of course, these direct job losses in the mining industry would lead to additional, indirect job losses in support and service industries. Assuming that each coal mining job produces two and a quarter indirect jobs, the national average, we estimate that the total employment loss due to a Senate-type bill would range from 130,000 to 195,000 jobs.

Other analyses support these conclusions. As an example, the Congressional Research Service analyzed employment impacts in the six states that I mentioned above. CRS chose those states because they are most affected by coal market shifts due to acid rain. The CRS study concluded that the direct coal mining job losses would range from about 22,000 to 33,000 in those six states. With the addition of the indirect job losses that would occur, the study indicates that between 71,000 and 107,000 people would lose their jobs.

The Energy Information Administration also has published a report analyzing the market impact of acid rain controls. EIA made specific projections on coal tonnage that would be lost in four states (Illinois, Indiana, Pennsylvania and Ohio).

Using EIA's projections, we calculated employment impacts based on 1982 average productivity in those states. The combined analysis showed the following: Illinois would lose 37 million tons of annual production by 1995, representing approximately 60 percent of total 1982 production. This would result in the loss of nearly 12,000 coal industry jobs in Illinois. Using again the 2.25 indirect jobs multiplier, the total job loss in Illinois would approach 39,000.

Indiana is projected by EIA to lose 14 million tons of coal, representing about 47 percent of 1982 production. This would mean a direct job loss of over 3,000 in the state's coal mining industry, while the total job loss would be near 10,000. Pennsylvania is estimated to lose 20 million tons, representing 27 percent of 1982 production. Coal mining job losses would reach almost 8,000 and total job losses for Pennsylvania would be over 25,000. EIA projected Ohio to lose 16 million tons or 44 percent of 1982 production. The coal mining industry in Ohio would lose over 5,500 jobs while, overall, Ohio would lose about 18,500 jobs.

Thus, in four important coal mining states, EIA projections of tonnage losses translate into over 28,000 coal mining jobs lost and a total loss of nearly 93,000 jobs. Clearly, these employment impacts are unacceptable. Moreover, the magnitude of the job losses points out the serious danger of adopting any acid rain control proposal that allows utilities to take actions that will create widespread unemployment among coal miners.

Because of the drastic economic and employment consequences threatened by a Senate-type acid rain bill, Representatives Waxman and Sikorski introduced H.R. 3400, which attempted to reduce sulfur dioxide and other pollutants without wholesale job disruption in coal mining communities. This was the first emissions reduction proposal that sought to protect jobs. Coal miners were encouraged that members of Congress finally were beginning to undertand that acid rain legislation posed a serious threat to their communities. Despite the good intentions, however, there were major problems with the House bill.

H.R. 3400 would require a total emission reduction of 10 million tons of sulfur dioxide and 4 million tons of nitrogen oxide in the 48 contiguous states by 1993. Approximately 6 million tons of the SO_2 reduction would be achieved by a mandatory retrofit of scrubbers on the fifty largest utility emitters, those with an emission rate above 3 pounds per million Btu.

Although a major purpose of H.R. 3400 was to prevent job losses among coal miners, our analysis indicated that it, too, would cause unemployment, but to a lesser degree than its Senate counterpart. These job issues could occur in two ways. First, there is the potential, perhaps even the likelihood, that a significant amount of fuel switching would occur as utilities attempt to meet the three-million ton residual sulfur dioxide reduction contemplated in the bill. Second, we believe that the age of some of the targetted "top fifty" utility plants would encourage the companies to retire some of the units instead of retrofitting them with scrubbers.

With regard to the additional three-million ton reduction requirement, we have examined potential job losses in the coal industry if fuel switching occurs. Our analysis shows that over 13,600 miners could lose their jobs in six coal producing states if 50 percent of the three-million ton reduction was achieved by fuel switching. Using the 2.25 jobs multiplier, we project a total job loss of about 44,000. For purposes of comparison, we reviewed tonnage disruption estimates for the same areas that were calculated by the Congressional Office of Technology Assessment (OTA) and Peabody Coal Company. Applying the same methodology to these estimates, we calculated coal miner job displacement. OTA's estimate of tonnage displacement results in slightly more than 9,700 miners losing their jobs while Peabody's estimate for those six states shows nearly 15,300 coal miners would lose their jobs. The total number of jobs displaced, of course, would be much higher, ranging from 31,000 to 49,000.

In addition to these potential job losses, we are concerned that the age of many of the power plants that would be targeted for scrubbers under H.R. 3400 will lead the utility companies to retire them early rather than apply scrubber technology. A review of specific utility in-service dates shows that nearly half of the units would be thirty years old by 1990, the date for compliance. In terms of unemployment, a plant that is shut down is just as devastating to coal miners as one that switches its fuel supply.

The UMWA believes that the magnitude of job losses described herein is unacceptable, particularly in light of the current, and growing, lack of scientific consensus on the issue. Legislation that would disrupt the economies of major coal producing regions, lead to unnecessary increases in oil and gas use, seriously hamper the coal expert trade, and lead to increased unemployment among coal miners simply cannot be justified.

UMWA coal miners always have had a strong interest in maintaining a clean and healthful environment. On the job, we are exposed to one of the most dangerous and unhealthful environments on earth. Because of this, we are adamant in our conviction that the clean air laws must be designed to protect and promote public health. Further, we strongly support an adequate margin of safety for our vulnerable citizens in the primary ambient air quality standards.

The debate on acid rain, however, has nothing to do with public health. It is a welfare-related issue and, therefore, must be viewed along with other important national goals. When Congress rewrote the Clean Air Act in 1977, it had several goals in mind concerning the coal industry. In pursuit of cleaner air, Congress expressly intended to:

- promote the development of all our coal resources;
- enhance employment opportunities; and
- give equal opportunity to all regions of the country to compete for economic development.

The UMWA agrees with and supports these objectives. Current suggestions for acid rain control, however, fly in the face of each of these goals.

There is no necessary adverse correlation between the use of coal and air quality. That should be clear from the advances that have been made in improving air quality over the past decade. From 1973 to 1982, total emissions of sulfur dioxide dropped 26 percent while emissions from electric utility plants declined by about 17 percent. The improvements are all the more impressive in light of the significant increases in coal utilization over the same period; total coal consumption grew 25 percent and utility coal use increased by 53 percent. Because of the concerns that all of us share, progress toward even better air quality will continue. Moreover, it will accelerate as new power plants replace those that are phased out of service. It would be extremely foolish, of course, to cease our vigilance. It would be even more foolish to look for a "quick fix" at the expense of this nation's coal miners and their communities.

Thank you.

CANADA'S ACID RAIN POLICY

Pierre Vachon

Senior LRTAP Manager
Priority Issues Directorate
Environment Canada
16th Floor, PZM Building
Hull, Quebec K1A 1C8
Canada

What I'd like to leave you with is a very brief review of what is happening in Canada, what our current situation is and where we are going. It may be useful to compare that to what has been happening in the United States.

Canada's policy on acid rain is to obtain reductions in sulfur dioxide emissions in order to eliminate damaging loadings to our environment. In February 1982, federal and provincial environment ministers agreed to an environmental objective: we are committed to limiting wet sulfate deposition to no more than 20 kilograms per hectare per year, which, according to our scientists, is the level needed to prevent damage to moderately sensitive lakes and rivers.

Allowable SO_2 emissions in eastern Canada in 1980 totalled about 4.5 million tons. The major source of emissions is the nickel and copper smelting industry which produced about 60 percent of SO_2 emissions. Utilities produce about 16 percent and non-utility fuel use about 13 percent. Any acid rain control program in Canada will have to focus primarily on the smelting industry but these other sources are also important.

We have agreed to reduce our SO_2 emissions in the eastern part of Canada by 50 percent by 1994. As a result of programs, regulations and commitments by provincial governments and the federal government, a 25 percent reduction in SO_2 emissions from 1980 base case levels will be in place by 1990. These reductions include:

1) Significant cutbacks in emissions at the INCO nickel smelter in Sudbury, Ontario, through processes involving increased pyrrhotite rejection;

2) A 40-percent reduction in SO_2 emissions from the copper smelter in Noranda, Quebec;

3) A 43-percent reduction in SO_2 emissions from Ontario Hydro, Canada's largest utility, through process changes such as coal washing, blending fuel and increased use of nuclear generated power;

4) A 300,000-ton reduction in SO_2 emissions as a result of switching from coal and oil to natural gas in many non-utility burners and upgrading of light and heavy oils.

A 50-percent reduction will mean that total annual emissions of SO_2 from the eastern part of our country after 1994 will be about 2.3 million tons. This level of emissions in turn means that many areas of Canada will receive less than 20 kilograms of wet sulfate per hectare per year.

We are now moving towards establishing better NO_x standards. We are in the process of reaching lower lead standards by the end of 1986 by reducing them to 0.29 grams per litre. In the field of research, the Canadian government, in cooperation with the provinces, is investing $1 per capita in research this year. As to the non-ferrous smelting industry, which is our main source of SO_2 pollutants, we are moving in the direction of modernizing that industry; by modernizing we will achieve greater productivity and clean up. We are introducing and implementing a policy of fuel switching that encourages home heating and industrial boilers to use cleaner fuel. With respect to reduction of SO_2 emissions, we are proceeding on schedule. If we take our base year of 1980 when emissions amounted to some 4 million tons, we have reduced in these four years by 15 percent. We are moving ahead with the program which has an interim goal of 25-percent cuts by 1990 and a final goal of 50 percent by 1994.

However, even if Canadian SO_2 emissions were to cease altogether, we could not protect all sensitive regions in Canada; deposition would still exceed 20 kg/ha/yr in some areas. This demonstrates the significance of the contribution of SO_2 from sources beyond our borders. More than 50 percent of Canada's acid rain problem originates in the United States. At the same time, 10 to 15 percent of the acid rain problem in the northeastern U.S. comes from Canada emissions.

You may ask why Canada is proceeding and the reason is that there is potential damage of immense importance to the Canadian economy. We estimate potential damage in tourism alone of 1.1 billion dollars; in terms of damage to forests, to our salmon rivers, to our lakes, to agriculture and structures, damage of some 15 billion dollars for a total of some 8 percent or more of our gross national product. We know that hundreds of our lakes are dead, thousands of our lakes are in peril, and that we want to prevent this situation from deteriorating. The question is, are we alone in suffering this damage, and the answer of course is no, and this leads me to the peculiarity of the damage caused by acid rain. It seems to me that it boils down to the basic fact that the nations which are located upwind reap the benefit of industrial production and that the nations which are downwind suffer the environmental and economic damage.

The cost of action is high but the cost of inaction is even greater and if we delay action, that cost may exponentially increase. We will never have a complete body of information and scientific knowledge that will determine conclusively what the causes are. But we can see the danger of irreversible damage and we want to prevent it. Water and air and soil, forests, structures and buildings are under stress, and ultimately, so is human health.

It is not surprising that public opinion surveys show that 8 out of 10 Canadians consider acid rain a serious problem.

Let me close quickly by making three points relevant to this conference. First, Canadian politicians have taken a risk-management approach and concluded that action was necessary now; they have made decisions on these convictions. Second, the Canadian approach is based on a wager that future technology developments will enable emission reductions to be achieved without causing unemployment; the future will tell if this was right. Third, there is a need for U.S. studies, particularly those based on cost-benefit and least societal cost analyses, to take account of the environmental costs borne by Canada. Study results presented here did not do so. Had they factored-in those consequences, I suggest the results would have clearly favored the need to take clean-up action in the U.S. and Canada to the benefit of the North American continent.

CONFERENCE SPEAKERS AND PARTICIPANTS

William G. Abbott
U.S. House of Representatives Staff

Jerry Ackerman
The Boston Globe

Elizabeth Agle
National Clean Air Coalition

James E. Anderson
Jersey Central Power & Light Co.

R. Assinoss
Environmental Action Foundation

James Atwater
Acid Rain Information Clearinghouse

Mrs. James Atwater

Lise Bacon
National Assembly, Canada

Paul C. Bailey
Southern Company Services, Inc.

David Baker
League of Women Voters

William Balson
Decision Focus, Inc.

Thomas W. Barlow
Florida Power & Light Company

Joseph R. Barse
U.S. Department of Agriculture

Patsy Y. Baynard
Florida Power Corp.

Loretta C. Beaumont
U.S. Department of Energy

Judy Bender
Newsday

Richard Bernknopf
U.S. Geological Survey

Carl Bernsten
Society of American Foresters

Rosina M. Bierbaum
U.S. Congress Office of Technology Assessment

W.D. Boone
Union Pacific Corp.

Constance Boris
University of Michigan

John C. Brennan
American Electric Power Service Corp.

Stephen Brick
Public Service Commission of Wisconsin

Mr. Brohan
American University, Washington, D.C.

Matt Brosius

Stuart S. Brown
Consolidation Coal Company

Susan M. Buffone
National Parks & Conservation Assn.

S.C. Burks
USI Press

Frederick G. Carlson
Pennsylvania Department of Environmental Resources

Douglas Carter
U.S. Department of Energy

Frank Caskins
C-Span

Lee Catalano
Power Magazine

Richard H. Chastain
Southern Company Services, Inc.

Tom Choman
Coal Outlook

John R. Cline
PEPLO Air Quality

Robert Coe
U.S. Department of State

Herman F. Cole, Jr.
Adirondack Park Agency

Greg Cooper
ANJEC

Daryl Cowell
Environment Canada

Scott H. Cragle
Pennsylvania Power & Light Co.

Robert W. Crandall
The Brookings Institution

Alex Cristofaro
U.S. Environmental Protection Agency

Thomas D. Crocker
University of Wyoming

Rae E. Cronmiller
National Rural Electric Cooperative Association

Elizabeth Crowley
North Carolina Dept. of Natural Resources and Community Dev.

J. William Currie
Battelle Pacific Northwest Laboratories

Edward P. Curtis, Jr.
Genesee Public Affairs

Nancy S. Dailey
Oak Ridge National Laboratory

Guy Darst
Associated Press

Malcolm Dole, Jr.
State of California Air Resources Board

Eric Dolin
National Wildlife Federation

Daniel J. Dudek
University of Massachusetts

Frederick C. Dunbar
National Economic Research Associates, Inc.

Clint Duncan
Central Washington University

Chris Dunsky

Ronald Dwight
Carnegie Endowment for International Peace

Carol Collyar Dykers
Charlotte Observer

John Dykers
Charlotte Observer

W.R. Effer
Ontario Hydro

Christian Elichegaray
Air Quality Agency, France

Charles Elkins
U.S. Environmental Protection Agency

David Elliott
Maine Office of Legislative Assistance

Elie Fallu
National Assembly, Canada

Susan F. Farrell
Edison Electric Institute

Howard J. Feldman
Middle South Services, Inc.

Robert Fenton
University of Winnipeg

W.C. Ferguson
INCO LIMITED

Kathy Ferland
Radian Corporation

Hans Foerstel
Environment Canada

Catherine M. Foley
Salt River Project

Gary Foley
U.S. Environmental Protection Agency

Katherine Foote
North Carolina Dept. of Natural Resource and Community Dev.

F. Frantisak
NORANDA, Inc.

G. Alexander Fraser
Canadian Forest Service

A. Myrick Freeman, III
Bowdoin College

James M. Friedman
Benesch, Friedlander, Coplan & Aronoff

David H. Fyock
General Public Utilities Service Corp.

Howard Gaines
New England Congressional Inst.

David A. Gansner
USDA Forest Service

A. Gilbert
University of Vermont

P.H. Giles
Environment Canada

T.J. Glauthier
Temple, Barker, Sloane, Inc.

Ruth Gonze
American Public Power Assoc.

Robin Gordon
League of Women Voters

Leon Green
Clean Coal Coalition

Norma Greenway
Canadian Press

Carl Griffith
Ontario Ministry of the Environment

James K. Grott
Southern Company Services, Inc.

James Groves
Kentucky Legislative Research Committee

Collot Guerard
Federal Trade Commission

Tom Haederle
Berns Bureau

Charles Hakkarinen
Electric Power Research Institute

O.F. Hall
Virginia Polytechnic Inst. & State Univ.

W.C. Hamilton
CONOCO Research & Development

Dennis Hammond
Florida Game & Freshwater Fish Comm.

Mr. Hamudi
American University, Washington, D.C.

Jim Hannah
Associated Press

Joe Harrison
Berns Bureau

David G. Hawkins
Natural Resources Defense Council, Inc.

Maureen L. Hellman
AMAX Coal Company

Ron Hemphill
NERCO, Inc.

Ross Hemphill
OARDA Hatch Project

Ray Herman
National Park Service

Owen W. Herrick
USDA Forest Service

R. Heuwinkel
Natl. Oceanic & Atmospheric Admin.

Edward L. Hillsman
Oak Ridge National Laboratory

Wallace D. Hoffman
Salt Lake Tribune

William Hogan
Harvard University

James F. Hornig
Dartmouth College

Betty Hudson
FLUOR Corporation

Jay S. Jacobson
Boyce Thompson Institute

Michael J. Jirousek
Cleveland Electric Illuminating Co.

Ronald Jonash
Arthur D. Little and Co.

Julian Josephson
Environmental Science & Technology

Bruce Jutzi
Canadian Embassy

Gerald J. Kabel
American Natural Resources

Robert Kaiser
Ohio Edison Co.

Bob Kane
U.S. Dept. of Energy

William G. Karis
Consolidation Coal Co.

Wendy Katz
National Environmental Development Assn.

Betty Katzner
Face-to-Face Program

William R. Kaul
Cooperative Power Assn.

James M. Kawecki
TRC Environmental Consultants

Lori Keesey
McGraw Hill "Coal Week"

Danford G. Kelley
Environmental Protection Service, Canada

Nancy Kete
John Hopkins University

Allen V. Kneese
Resources for the Future

Kris W. Knudsen
Duke Power Co.

Mr. Koff
American University, Washington, D.C.

Lorelli Kornell
Prentice-Hall

Peter J. Kuch
U.S. Environmental Protection Agency

Barbara Kwetz-Allan
Massachusetts Executive Office of Environmental Affairs

Frederick J. Lange
Niagara Mohawk Power Corp.

David L. Lantz
Hoosiers for Economic Development

Thomas S. Lareau
U.S. Environmental Protection Agency

Giles Laterriere
Government of Quebec

Douglas A. Latimer
Systems Application

John A. Laurmann
Gas Research Institute

Lester B. Lave
Carnegie Mellon University

Raymond List
Norton Hambleton

Susan List
Norton Hambleton

K.H. Ludum
Texaco, Inc.

Thomas A. Lynch
Florida Dept. of Environmental Regulation

James R. Lyons
Society of American Foresters

Molly Macauley
Resources for the Future

Paul W. MacAvoy
University of Rochester

Conrad MacKerron
BNA Environment Reporter

Susan Macy
Genesee Public Affairs

Paulette Mandelbaum
Chemical Engineering and Policy Analyses (CEPA)

C. Robert Manor
PEPLO Air Quality

William E. Marlatt
Colorado State University

Bill Matthews
Thomas Newspapers

Janet G. McCabe
Massachusetts Dept. of Attorney General

Julie P. McCahill
Mead Corporation

Frank J. McDowell
Public Service of Indiana

Robert H. McFadden
Motor Vehicle Manufacturers Assn.

Charles W. McGrady
Georgia Chapter, Sierra Club

Ira E. McKeever
W.R. Grace & Company

Patrick H. McNamara
Dow Chemical Company

Clifton Metzner, Jr.
U.S. Department of State

Denise Michel
U.S. Senate Staff

Paul A. Miller
Rochester Institute of Technology

Robert Missen
Pacific Power & Light Co.

Curtis D. Moore
U.S. Senate Staff

Ronald J. Nesse
Battelle Pacific Northwest Laboratory

A. Lee Nesslage
Acid Rain Information Clearinghouse

John Nester
Berns Bureau

Joanne Nichols
Massachusetts Dept. of Environmental Management

David C. Nicholson
Weyerhaeuser Co.

D. Warner North
Decision Focus Inc.

Jack O'Donnell
Advanced Energy Dynamics, Inc.

Craig Oliver
Canadian Television

Dana Orwick
Aspen Institute

Larry B. Parker
Library of Congress

S.L. Pernick, Jr.
Duquesne Light Co.

Sueanne Pfifferling
U.S. Senate Staff

Stephanie Pfirman
U.S.House of Representatives Staff

Paul R. Portney
Resources for the Future

David Powell
Batelle Pacific Northwest Laboratory

Allan G. Pulsipher
Tennessee Valley Authority

Sally Rand
U.S. House of Representatives Staff

Richard Rao
EBASCO Services, Inc.

John O. Rawlings
North Carolina State University

Philip Reed
Environmental Law Reporter

Wallace Reed
Virginia Air Pollution Board

James L. Regens
University of Georgia

John Reilly
Institute for Engineering Analysis

Michael Richards
CBS Radio

Richard Richels
Electric Power Research Institute

Jacques-Andre Rivard
Canadian Broadcasting Corp., National French News

Ray Rivers
Dept. of Fisheries & Oceans, Canada

Marilynne K. Roberts
Project Environment Foundation, Minnesota

W.L. Roberts
Martin Marietta Energy Systems

Randy A. Roig
Maryland Energy Administration

Mike Ruby
Sierra Club

Milton Russell
U.S. Environmental Protection Agency

Joni M. Sage
Group W Newsfeed

Ajay Sanghi
New York State Energy Office

Jean-Louis Sasseville
Institute National de la Recherche Scientifique, Quebec

Claude Sauve
Quebec Min. de L'Environnement

Paul B. Schumann
UCLA; National Research Council

Seth Schwartz
Energy Venture Analysis, Inc.

Roger A. Sedjo
Resources for the Future

James Seligman
U.S. House of Representatives Staff

Curtis Seltzer
Southern Appalachian Leadership Training

David C. Shadle
Jersey Central Power & Light Co.

David J. Shaw
New York State Dept. of Environmental Conservation

R.W. Shaw
Atmospheric Environment Service

Robin A. Skitt
Peabody Holding Co.

Frank Skutta
General Electric

James E. Slater, Jr.
Bechtel Power Corp.

Martin Smith
Office of Policy Development, White House

Hoff Stauffer
ICF, Inc.

Harry L. Storey
A.C.E.

Frederick Stoss
Acid Rain Information Clearinghouse

Anne D. Stubbs
Coalition of Northeastern Governors

Mary Stultz
Acid Rain Information Clearinghouse

Susan Subak
Carnegie Endowment for International Peace

Jim Sweeney
Air/Water Pollution Report

Charles Tasuordi
Associated Press

William B. Taylor, Jr.
U.S. Senate Staff

Elizabeth Thorndike
Acid Rain Information Clearinghouse

George Tomlinson
ARIC Advisory Board

Ian Torrens
Organization for Economic Cooperation & Development

Richard Trumka
United Mine Workers

Jacquelyn L. Tuxill
New Hampshire Citizens' Task Force on Acid Rain

Thomas A. Ulasewicz
Adirondack Park Agency

Myron Uman
National Academy of Sciences

Pierre Vachon
Environment Canada

Alan VanArsdale
Massachusetts Dept. of Quality Engineering

Jan vanKaenel
Associated Press

Herbert B. Visscher
Dames & Moore

William Wagner
Acid Rain Information Clearinghouse

William D. Watson
U.S. Geological Survey

John K. Weglo
Falconbridge, Ltd.

Richard A. Wegman
Wellford, Wegman, Krulwich, Gold & Hoff

William R. Wilkes
Monsanto Research Corp., Mound Laboratory

Carol Willey
Acid Rain Information Clearinghouse

Henry G. Williams
New York State Dept. of Environmental Conservation

Jeff Williams
Rochester Gas & Electric

Richard Wilson
Harvard University

Derek Winstanley
National Acid Precipitation Assessment Program

Robert E. Wood
Montana-Dakota Utilities Co.

W.R. Woodall, Jr.
Georgia Power Co.

Les Woodruff
CBS Radio

David R. Wooley
New York State Department of Law

M.G. Wrobel
Canada Library of Parliament

Kenji Yamada
Gifu College of Economics, Japan

Charlotte Zieve
University of Wisconsin

INDEX